*UNDERSTANDING CATASTROPHE*

The Darwin College Lectures

# Understanding Catastrophe

*EDITED BY JANINE BOURRIAU*

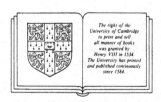

The right of the
University of Cambridge
to print and sell
all manner of books
was granted by
Henry VIII in 1534.
The University has printed
and published continuously
since 1584.

CAMBRIDGE UNIVERSITY PRESS

CAMBRIDGE

NEW YORK    PORT CHESTER

MELBOURNE    SYDNEY

CAMBRIDGE UNIVERSITY PRESS
Cambridge, New York, Melbourne, Madrid, Cape Town, Singapore, São Paulo

Cambridge University Press
The Edinburgh Building, Cambridge CB2 2RU, UK

Published in the United States of America by Cambridge University Press, New York

www.cambridge.org
Information on this title: www.cambridge.org/9780521413244

First published 1992
This digitally printed first paperback version 2006

*A catalogue record for this publication is available from the British Library*

ISBN-13  978-0-521-41324-4 hardback
ISBN-10  0-521-41324-9 hardback

ISBN-13  978-0-521-03219-3 paperback
ISBN-10  0-521-03219-9 paperback

# CONTENTS

# Introduction
# Understanding catastrophe

*GEOFFREY LLOYD*

How, in a world of ever-increasing specialisation, can effective communication be maintained? This is not just a matter of communication *between* different fields of technical knowledge, but also often *within* them. It is no longer merely the old problem of the two cultures and of the gap that separates the arts and the sciences. Within each of the broad areas into which the natural sciences, the social sciences and the humanities are customarily divided, there are considerable obstacles to the exchange of ideas and information. It is not just biologists and physicists, but also mathematicians who may find the work of other mathematicians inscrutable. It is not just to scientists that the technical language of philosophers, literary critics, and historians, may seem just so much jargon.

Efforts at cross-disciplinary communication often sadly amount to forlorn attempts at popularisation, where much of the technical content is drastically over-simplified or omitted altogether. Yet opportunities for significant cross-disciplinary exchange exist, notably where several different fields make use of the same or similar key concepts or methods – or where debate focuses on similar problems or exhibits cognate preoccupations. Here, when specialists in each of several areas are prepared to pool ideas, new insights may emerge and the debate may be advanced when the issues are seen in a fresh multi-disciplinary perspective.

In 1986 Darwin College, Cambridge, inaugurated a lecture series that aimed to promote the inter-disciplinary exploration of fundamental concepts. World experts from a variety of specialisations are invited to participate first in a series of public lectures, then in a joint publication in which they tackle issues of great contemporary importance. Each contributor is given a free hand to discuss those aspects of the general theme that appear most interesting or problematic. Following successful series devoted to *Origins, The Fragile Environment*, and *Ways of Communicating*, the topic chosen for the 1990 series was *Understanding Catastrophe*.

Catastrophe is a theme that offers particularly exciting prospects for cross-disciplinary exploration. The concept is used in a variety of areas in the natural sciences, in mathematics, in archaeology, in history, and in ecology especially. In the sequence of detailed studies that follow, we move broadly from the cosmological to the human.

Robert Kirshner deals with the massive astronomic catastrophe represented by the phenomenon of the supernova – the sudden collapse of the dense stellar core leading to the formation of a neutron star. We are here brought right up to date with the latest astronomical observations and experiments and with the latest theories both on supernovae themselves and on their cosmological implications.

Walter Alvarez and Frank Asaro track down possible reasons for the extinction of the dinosaurs, exploring the hypothesis that this may have been due to the catastrophic impact of extra-terrestrial objects. However, they also consider alternative hypotheses, including the possible effects of volcanism, and here too up-to-the-minute observations are brought to bear on the problems.

With Martin Rudwick, we turn to the origins of the geological debate in Darwin's day. Then, the mainstream of geological theory represented by the catastrophism of Whewell and Sedgwick was challenged by Lyell's *Principles of Geology*, a classic text for what came to be dubbed uniformitarianism, the view that natural changes can be adequately explained by postulating continual gradual alterations. As Rudwick shows, the outcome of this debate was not just a matter of the scientific data and arguments presented on either side, but also depended on the personalities, the prestige, and the persuasive talents of the debaters.

Christopher Zeeman in Chapter 4 explores the mathematics of catastrophe theory, and for good measure takes as his example Darwinian evolution. Here we learn that discontinuities in time, space and form are not necessarily the product of sharp discontinuities in the causal factors involved. Zeeman's argument is that such changes may be not merely consistent with the hypothesis of continuous causes, but may even be the consequence of those hypotheses.

Claudio Vita-Finzi studies earthquakes, both major and minor, and points out both their frequency and the variability of their effects. Here, and in the chapter that follows by Nicholas Cook on storms and cyclones, the question of the human response to catastrophes is explored. What planning measures are appropriate to cope with rare but predictable catastrophic events?

In Chapter 7 Peter Garnsey takes the issue of human response one stage further when he defines famine partly in terms of the breakdown of the social, political and economic order that drastic food shortage may bring about. He significantly remarks on the avoidability of many such a breakdown.

Finally, Roy Porter discusses consumption, both the disease and the economic activity now associated with consumerism. He shows how the disease has been used in different periods from the seventeenth century to the present day as a metaphor for evil, and how different attitudes towards it can be used to reveal fundamental preoccupations of both the medical profession and the society it serves.

What emerges from these wide-ranging topics is first the great variety of phenomena to which the notion of catastrophism can be applied – from those of cosmic proportions to the collapse of buildings. Second, the diversity of human response, whether practical or imaginative, gives food for thought. Third, the debate on the explanation of catastrophic events provides a focus for the investigation of changes in attitudes to the world and of the relationship of humans to it.

In these studies, world experts grapple with the fundamental problems, explore alternative hypotheses, bring to bear the newest evidence, and probe the limits of our understanding. The controversy between catastrophists and uniformitarians was at the centre of scientific debate in the nineteenth century. This book is testimony to its ongoing vitality and interest.

# Supernovae and stellar catastrophe

*ROBERT P KIRSHNER*

A series of catastrophes has brought each of us to our present state. Author and reader, the calcium and iron atoms that form our bones and blood were forged in the crucibles of stellar catastrophes – supernova explosions that took place 5 thousand million years ago. The death of stars leads to the chemical enrichment of the universe, its growth in complexity, and more particularly to the oxygen we breathe and the carbon atoms that ink the pages of this book.

## STRANGER THAN FICTION

This cosmic benefaction results from the destruction of a massive star in a remarkable astronomical phenomenon where, for a few weeks, a single star shines as brightly as 10 thousand million suns, ejecting the interior of the supernova into the interstellar gas. New stars, planets, and books can be formed out of this richer mixture. In this sense, we are all made of stardust.

A supernova explosion is the sudden death of a star that results from the violent collapse of its core into a dense neutron star: an object with the mass of a star, but the size of a city. The energy released by this precipitous contraction powers the supernova, blasting the star apart, cooking its interior into new elements, and heating its atmosphere until it shines as brightly as a galaxy.

The constancy of stars is a literary cliché, but it is misleading. A squirrel may regard a redwood tree as eternal, but that is just a matter of the life span of the observer. While it is true that even short-lived stars endure for millions of years, 10 000 human generations, we know that stars change. The biography of a star from birth, through the crises of adolescence, into a productive middle age, and finally to a catastrophic end is called 'stellar evolution' by astronomers even though it is just the development of a single star. Interactions between particles that are smaller than an atom shape the events on a stellar scale and change the contents of the universe.

There is real evolution in the universe and this provides the context for supernovae: stars are found in galaxies, which have endured for 1000 generations of massive stars. The motions of the galaxies indicate that the pageant of stellar evolution is played out against a larger background of an expanding universe. The evidence is that the universe we see today evolved from a hot dense phase about 10 to 20 thousand million years ago: the Big Bang. The overall expansion and the development of structures that matter takes today, such as stars, galaxies, and the uneven distribution of galaxies in huge clusters, sheets and voids is one kind of evolution. There is also real evolution in the development of structure on the atomic scale. Stellar events take the simple forms of matter left over from the Big Bang, hydrogen and helium, and change them into more complex atoms that yield the richness of the world around us. Supernovae play the key role in the transmogrification of matter and the progressive enrichment of the Universe in heavy elements.

More than one route leads to stellar catastrophe. Here I emphasise the death of massive stars, some 20 times more massive than our sun. Stars of just a few solar masses can also meet with a violent end as supernovae, but lightweights require an accomplice to achieve this brilliant end. The tale of nuclear change and stellar collapse for massive stars is sufficiently intricate to illustrate the main ideas and, more importantly, has been subjected to an illuminating observational test, not by brilliant experimental design, but by great good fortune. Supernova 1987A, the brighest supernova since Tycho's of 1572 and

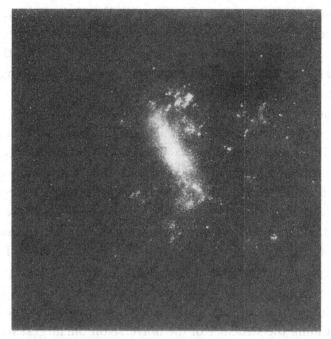

*Figure 1* The Large Magellanic Cloud. The Large Magellanic Cloud is a small irregular galaxy that is a satellite of our own Milky Way Galaxy at a distance of about 150 000 light years. It has active regions of star formation where young, massive stars traverse the complete path of stellar fusion and explode in the regions where they formed. Supernova 1987A was the result of such a massive star in a region of star formation.

Kepler's of 1604, was discovered just three years ago in the nearest galaxy to our own Milky Way: the Large Magellanic Cloud (Figure 1).

## THE LARGE MAGELLANIC CLOUD

Our sun is one of 100 thousand million stars in the Milky Way Galaxy, a great spinning spiral galaxy. It takes light about nine minutes to reach the earth from the Sun, and a few years for light from the Sun to reach the nearest stars. For convenience, astronomers use the term light year (LY) to indicate the distance light travels in a year. This device also has the salutary effect of reminding us that events we observe today with light gathered by telescopes are the result of events that took place in the past. We see the light from the Sun as the Sun was nine minutes ago, and from nearby stars as they were a few years ago. The largest dimension in the universe would be the distance that light has travelled in the time since the Big Bang, corresponding to 10–20 thousand million LY.

The diameter of our galaxy is about 100 000 LY and the typical separation of galaxies is a few million light years. Galaxies, like grapes, cluster together on scales which are 10–100 times larger, and the most extended voids and lumps in the cosmic distribution of matter are larger still. We do not yet know how large the largest structures in the universe are. We expect them to be much smaller than the 10–20 thousand million LY that characterises the size of the universe, but our current efforts at cosmic mapping have not taken us very far into those deep waters. The lives of stars hinge on details of how protons and neutrons combine. Similarly, current thinking is that the inventory and interactions of subatomic particles may also hold the key to understanding the structure of the universe on the largest scales.

Nearby structures in the universe were noted by Magellan in his 1521 world tour. Observers in southern latitudes can easily pick out the two conspicuous fuzzy patches of light we call the Large and Small Magellanic Clouds. The Large Magellanic Cloud (LMC), the site of SN 1987A, is a small galaxy of stars and gas at a distance of about 160 000 LY. Compared to the 100 000 LY span of our own galaxy, this is truly in our neighbourhood and SN 1987A was easily accessible to the whole range of modern instruments that were not available to Tycho, Kepler or any of our earlier colleagues. We are extremely fortunate to have seen this supernova in our own lives. If the LMC were just a little farther away, the light would not yet have reached us; if it were a little closer, our grandfathers would have seen the event.

The light from SN 1987A was first detected at the Las Campanas Observatory in Chile on 23 February 1987. Telescope operator Oscar Duhalde went out for a look at the sky while he put a kettle on to boil water for coffee and noticed something odd about the LMC. A few hours later, observer Ian Shelton walked over to the dome where

*Figure 2* Supernova 1987A observed on 25 February 1987.

Duhalde was working and reported he'd found a new star in the LMC. The object erupted near the part of the LMC known as the Tarantula Nebula (based on the local fauna) (Figure 3). We recognise this cloud of gas and dust as a hotbed of star birth. For the last 10 million years or so, there has been a lively baby boom of stars in this neighbourhood, leading in the end to a conspicuous bang.

## STARSHINE

A star is a ball of gas held together by its own gravitation. The gas is hot enough so that gas pressure supports the star and a stable balance is reached. The problem for understanding stars is that energy is

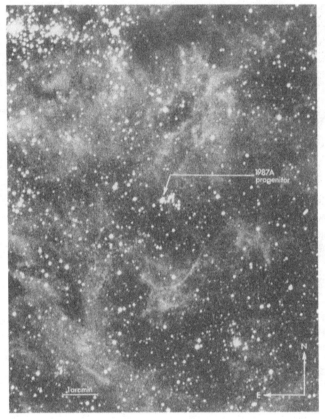

*Figure 3* The site of Supernova 1987A. Measurements made
after the outburst show that SN 1987A is coincident with the
massive star Sk -69 202 catalogued by Sanduleak. This is the
most direct evidence that supernovae come from massive stars.

escaping from the surface in the form of light. Where does the energy
come from to sustain that luminosity for the lifetime of a star? This
problem became very serious in the early decades of this century
when the age of the Earth was determined to be about 5 thousand million years, based on the accumulation of radioactive decay products.
Previous ideas that the Sun's energy source might be meteor infall or
(more plausibly) gravitational contraction could only account for the
Sun's output for about 20 million years. Fortunately, the same studies
of radioactivity that established the age of the Earth also gave an
inkling of the power source for the Sun. We now know that the Sun

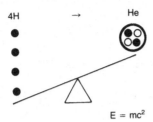

*Figure 4* Energy from fusion. Nuclear energy generation in stars results from the conversion of mass to energy according to Einstein's often quoted equation $E = mc^2$. Through a chain of nuclear reactions, four hydrogen nuclei are fused to form a helium nucleus. The mass of the ingredients exceeds the mass of the product and the balance shows up as energy. Since stars are composed of hydrogen, and the energy yield is large, nuclear fusion can power stars of the Sun's mass for more than 10 billion years.

gets its energy through nuclear fusion that takes place in the centre of the star. Through a chain of reactions, nuclei of the element hydrogen are converted into nuclei of helium. Since hydrogen is the most abundant element in the universe, each star contains its own powerful fuel supply.

This nuclear transmogrification has a very interesting side effect. The four hydrogen nuclei that go into the fabrication of a helium nucleus have a little more mass than the resulting helium nucleus. The small mass difference corresponds to the release of an amount of energy according to Einstein's well-known formula $E = mc^2$ (Figure 4). This formula is not just an icon of scientific inscrutability. It is a real equation that tells how much energy is released when a mass $m$ disappears. Since the mass is multiplied by $c^2$, where $c$ is the speed of light, and $c^2$ is a very large number, the yield of nuclear energy is a million times larger than the energy you get by taking the same amount of hydrogen and burning it by ordinary chemical reactions. Explosive effects of this immense energy release form the basis for nuclear weapons. A more beneficial result of nuclear fusion is that stars have a vast store of energy and low mass stars like the Sun can sustain a constant output of light for 10 thousand million years. This is good news for biological evolution. It is only recently that biological

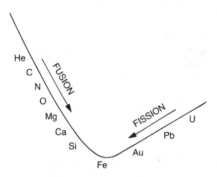

NUCLEAR BINDING

*Figure 5* Fusion and fission. Nuclear energy can be extracted by building up more complex nuclei. Hydrogen yields energy in fusing to make helium, helium can be fused in massive stars to make carbon, and other fusion products can be recycled to provide energy for the late stages of a massive star up to iron. Nuclei which have more neutrons and protons than iron yield energy when they fission into smaller fragments. Because iron lies at the bottom of this valley of stability, nuclear energy generation from fusion ends at iron. Synthesis of elements beyond iron takes place in supernova explosions.

evolution has produced anything which understands how long it took to get here.

Stars that are more massive than the Sun, like the one that formed 10 million years ago near the Tarantula Nebula, can perform even more remarkable changes in their composition. More massive stars have to support themselves with higher pressure at the centre, and for a gas that requires a higher temperature. The ferocious collisions in the fearsomely hot centre of a massive star can recycle the wastes of one burning stage into the fuel for the next. Higher temperatures allow the helium ashes of hydrogen fusion to become the fuel for the next stage – fusing helium nuclei three at a time to make carbon. Once again, the product has less mass than the ingredients and energy is released.

In a massive star, further evolution comes from ignition of successively heavier nuclei. Carbon is burned into oxygen, oxygen into silicon, and silicon into iron (Figure 5). Each of these burning stages takes place at a higher temperature, has a briefer duration, and yields a smaller return of energy on the ingredients used up. The history of stellar energy for a massive star is a progressive slide down the chain of nuclear fusion, locking up the matter into heavier and heavier nuclei in return for the energy extracted.

The structure of a massive star reflects this history. The outside is unburned material, the raw hydrogen that makes up over 90% of the elements we see. The next ring in is the helium created from hydrogen fusion. Inside the helium we find a zone of carbon and oxygen, then silicon and finally iron in the very centre. A star with an iron core is a star poised on the brink of catastrophe.

## ON THE BRINK

Massive stars die in a sudden, violent way because of the details of nuclear structure. Energy is extracted as a star fuses light elements into heavier ones up to the element iron. But iron is the most tightly bound nucleus. It is the nuclear turnip from which no blood can be squeezed. It takes energy to produce elements with more neutrons and protons than in an iron nucleus. Conversely, nuclei with more neutrons and protons, such as uranium, yield energy by fission: by splitting into smaller parts. We know that uranium exists, but a star cannot generate energy by synthesising uranium from iron.

After a star fuses the elements near silicon into iron in its core, it contracts and heats. In every previous stage this resulted in the ignition of a new fuel and produced a new era of stability. This time, it leads to disaster since all the available nuclear energy has already been extracted when the star formed iron. If you heat an iron core, you do not ignite a new energy supply. Worse, iron may begin to disintegrate, sapping energy from the core, leading to a sudden collapse.

A massive star collapses faster than a Florida Savings and Loan. The inner core has the mass of the sun and, drawn in by its own powerful gravitation, it shrinks from the size of the earth to about 10 km radius in one second. The star falls in at speeds approaching the speed of light until the inner core reaches the density of an atomic nucleus. The force that brakes the collapse is the strong nuclear force: the same force that binds neutrons and protons in atomic nuclei. Only here it is acting among $10^{56}$ particles, not just the 56 in an iron nucleus. When the collapse halts, a violent rebound sends a powerful pressure wave screaming out through the atmosphere of the star, heating it and blasting the outside of the star into interstellar space. Imagine a train slamming into a brick wall – when the engine stops, a violent wave travels back to the caboose.

As the pressure wave traverses the star, it heats some of the unburned material and cooks it to iron and beyond by bombarding it with protons and neutrons. A shock wave in a supernova achieves the ancient alchemist's dream of making gold out of iron, and goes on to his nightmare of changing gold to lead. It creates elements such as platinum and rubidium and mercury and tin that do not result from the energy generating fusion in stars. These elements are very rare in the universe, about $10^6$ times less abundant than the iron seeds from which they grow. Even in the enriched material of our own sun, only 2% of the mass is in elements beyond helium. The fact that we can sometimes find lumps of gold means that the process of forming a planet includes ways of concentrating these cosmic hen's teeth.

When the pressure wave arrives at the outside of the star, it rips off the outside of the star at speeds up to 10% of the speed of light, and heats it to glow as brightly as 10 thousand million suns. This is the brilliant explosion we recognise as a supernova.

However, most of the energy from this catastrophe does not go into nuclear change or motion or light. Deep in the centre of the star, just outside the forming neutron star, temperatures reach $10^{10}$ K. Since reliable testimony indicates that brimstone melts at a few times $10^3$ K, the centre of a supernova is 10 million times hotter than Hell! In these extraordinary conditions, most of the energy goes into the production of massless, chargeless particles called neutrinos. Neutrinos do not interact with matter by the electric forces that keep your feet from going through the floor. Neutrinos can pass through lead bricks, or even the earth, like light through a window. In a supernova explosion, calculations show that 99% of the energy released by the collapse to a neutron star is emitted in this ghostly form. Only about 1% rips apart the rest of the star, and only 0.01% is enough to produce the brightest optical display in the universe.

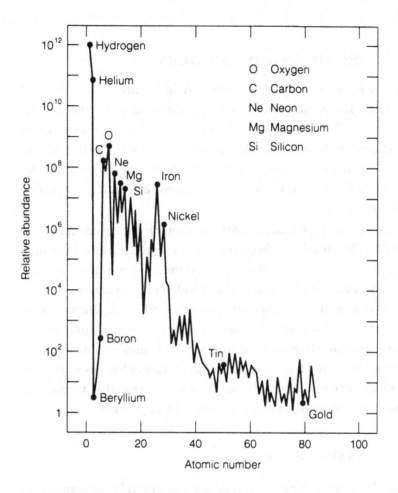

*Figure 6* Abundances of the elements. The pattern of chemical abundances observed in the solar system, the Sun, and in other stars reflects the origin of elements in the Big Bang (for helium) and in stars, notably supernovae, for all other elements. Note that the universe is mostly hydrogen and helium, with just a few per cent of the heavier elements. Of these, the nuclei produced in stellar burning stages are the most abundant. Note that iron and its neighbours are a million times more abundant in the universe than the rare heavier elements germinated from iron seeds in supernova shock waves.

## CHECKING THOSE FIGGERS

If this story were science fiction, we would find it implausible. It could only be fact. Massive stars with self-regulating nuclear reactors fusing the material of life, then having their cores crushed into single monster nuclei while blasting freshly synthesised gold into interstellar space are extravagant enough, but emitting 99% of the energy in an undetectable form is too fantastic a picture for any sane imagination to produce.

The distinguishing feature of natural science is that predictions and models (or wildly implausible stories) can be tested by observation. In the case of SN 1987A, we have employed the full arsenal of observational tools to see whether this elaborate idea of a supernova is really correct. We have advantages over Kepler's 1604 work. We have telescopes with sensitive light detectors, satellites above the atmosphere, neutrino detectors burrowed into the earth, and a much deeper understanding of the nature of light and matter to help us make and understand measurements of the supernova. Armed with modern tools, we can check the modern picture of a supernova.

## MASSIVE STAR?

Because the LMC is nearby, it has been intensively surveyed. One interesting list of 1400 bright, massive stars was compiled at Cerro Tololo Inter-American Observatory in Chile, by a mild-mannered Cleveland astronomer named Nicholas Sanduleak. SN 1987A turns out to be exactly at the same location as star number 202 on his list in the band 69° south of the equator. The local newspaper surprised the scientific community in February 1987 by reporting this identification with the headline 'Sanduleak explodes!'

Most supernovae are detected in much more distant galaxies, and it has never been possible to identify the progenitor star directly. While circumstances, such as being found in regions where massive stars are common, have suggested that supernovae come from massive

*Figure 7* Ol' Doc Dabble. The earliest publication of the neutron star model for supernovae in January 1934. The 'shrivelling' of stars to 'little spheres 14 miles thick' was suggested by Fritz Zwicky and Walter Baade in more formal scientific publications as well, just two years after the discovery of the neutron. This cartoon also suggests a sound approach to supernova studies: allowing someone else to find the supernovae coupled with a healthy regard for observational tests of theoretical ideas.

*Figure 8* Fritz Zwicky. Highly opinionated and creative prophet of supernovae, Fritz Zwicky worked at Caltech from 1925 until his death in 1973. Zwicky emphasised the importance of supernovae, proposing that supernovae are powered by gravitational collapse to a neutron star. He also suggested that clusters of galaxies are bound together by large amounts of unseen matter and that galaxies can act as gravitational lenses to amplify the light of distant objects – two areas of great interest in modern astrophysics.

stars, SN 1987A is the first supernova which is known beyond a reasonable doubt to be the result of exploding a particular massive star. The case is strengthened now, a few years after the explosion, by the fact that the blue supergiant star Sanduleak -69 202 is missing, exactly at the site of SN 1987A (see Figure 3).

While this star was on the list of massive stars in the LMC, there was nothing peculiar noted at the surface that gave any clue to the violent events brewing in the stellar core. The atmospheres of stars are so large that they respond slowly to rapid changes down below. So as the last few days of the star ticked away, while silicon burning went at a ferocious pace in the centre, and the core crept ever closer to a precipitous collapse, the outside of the star presented the same aspect that it had in Magellan's time and for 10 000 years before. We have gathered some clues from ultraviolet instruments on a satellite that Sanduleak -69 202 was not always a blue supergiant of the type Sanduleak catalogued, but has a more lurid past 30 000 years ago as a red supergiant, losing much of the star's hydrogen envelope in a sustained stellar wind.

## NEUTRINO EMISSION?

The central event in this type of supernova is the sudden collapse of the dense stellar core to become a neutron star. Just as water flowing downhill releases energy that can be used to run a mill or generate electricity, the downhill collapse of a stellar centre provides a sudden burst of energy that could lead to the emission of copious neutrinos and the disintegration of the star. But did this really happen? Neutrinos are hard to detect, but, just as you sometimes see the unexpected in the corner of your eye, astronomers' cousins in high energy physics found neutrinos from SN 1987A, while searching for the decay of protons.

The tremendous variety of matter we see around us is made of atoms. We now recognise all the elements as over 100 different arrangements of just a few basic particles: protons and neutrons in the atomic nucleus with an electrically balancing cloud of electrons. Detailed study of the structure of matter on smaller scales is the domain of particle physics and has revealed many additional particles, like the neutrino, and shown that the 'elementary particles' are themselves made of smaller constituents, the quarks. The aim in this enterprise is to understand the particles and the forces between them in the simplest way. One very popular way has been to see all the forces in nature as different aspects of a single force. These 'Grand Unified Theories' (GUTs to the irreverent) make one clear prediction: the lifetime of the proton should be finite. The consequences of this are startling: eventually all the matter we see making up stars and galaxies would evaporate. This doesn't correspond too well with everyday experience (where only money, keys, and the other sock are observed to evaporate) but that is only because the predicted lifetime is very long: about $10^{23}$ times the age of the universe.

Devising a test for this theory is challenging, but not impossible. The requirement is to gather enough protons so that you have several decays per year, and to think of a scheme that will enable you to detect the disintegration of a single particle. Water provides a cheap, transparent, easily handled source of protons. Six thousand tons is enough. If you hollow out a cavern in a mine and line the walls with light detectors you can test the picture for proton decay by waiting. You could fill in the quiet hours by composing a modest acceptance speech for the Nobel prize. Nature, however, had a surprise for the proton decay experimenters. They did not see the proton decay, causing several theoretical particle physicists to fall on their fountain pens in ritual suicide.

But nature had yet another surprise in store. The blast of neutrinos from the core collapse in Sanduleak -69 202 arrived at the Earth between the time that the experimenters announced they had not found any decaying protons and the moment of decision on future funding for the enterprise. The same detectors were excellent devices for catching neutrinos from the star's collapse: perhaps the best result from the search for proton decay.

On 23 February, at 7:36 UT, 1987 about $10^{10}$ neutrinos from the core of the collapsing star passed through every square centimeter of the earth. In the giant volume of the IMB detector in a salt mine in Ohio, eight were detected. At the same time, another 12 were snared at the Kamioka mine in Japan. The results agree well with the predictions for core collapse: they have the right number, the right energy and the right duration to be the result of emission from a neutron star forming in the LMC. Even better, the moment of core collapse agrees very well with the eyewitness accounts by the observers who discovered SN 1987A. The moment of the core collapse came just a few hours before the brightening of the star, just as it should if a shock wave carries the energy from the bouncing core to the surface.

## NEUTRON STAR?

Even before this stunning confirmation of the supernova picture, there were good reasons to believe the story based on the discovery of pulsars. Pulsars are spinning magnetic neutron stars and have been found at the sites of supernovae in our own Milky Way galaxy, most notably in the Crab Nebula. Another pulsar has been detected in a remnant in the LMC with the charming name of 0540-69.3 (The Digit Nebula?). Our observations of 0540 show that it is about 750 years old, and may well have been the previous supernova in the LMC. But it is one thing to find a corpse and another actually to see the smoking gun. In the case of SN 1987A, the direct observation of the neutrino pulse seems the best evidence for the formation of a neutron star. Unfortunately, although we have seen the flash of the gun, we have yet to find the corpse!

One of the most remarkable features of ordinary matter is that solid objects, like a lump of lead, are mostly empty space. While the mass of ordinary matter is concentrated in the atomic nucleus, most of the volume provides room for the electrons to orbit. In the case of a neutron star, the matter is compressed to the density of a nucleus, which means the individual neutrons are cheek by jowl. The increase in density is astonishing. While ordinary lead has a density that is a few times the density of water, neutron star matter is 100 million million times denser. A teaspoon full of neutron star stuff would have the mass of Mount Everest. A solar mass worth of matter is crammed into an object the size of a city.

For SN 1987A, there should be a city-sized clinker of neutron-star matter in the middle of the exploded star. Despite some early reports, the neutron star has not yet been found. If the neutron star is a spinning magnet, it may lead to flashes of light, as seen in other pulsars. If the neutron star does not have a strong magnetic field, it might be revealed by the emission from matter falling back on the neutron star. In any case, this part of the story remains a loose end. One possibility is that the neutron star may have collapsed to an even denser state and become a black hole, a region of space so dense that not even light can escape. Black holes are hard to detect, but we might see emission from matter falling into its gravitational grip.

## EXPLOSION?

When the core collapsed in Sanduleak -69 202, forming the neutron star and emitting the neutrino blast, a small fraction of the energy coupled to the atmosphere of the star. But the pocket change of millionaires can be very significant. In this case, one per cent of the energy of the supernova is more than enough to heat and destroy the rest of the star. Measurements made just after the first detection of the supernova showed that the outer layers of the star were ejected at one tenth the speed of light. Even gas from deep within the star was blown out, including the heavy elements synthesised in the long life of the star and the freshly synthesised elements that resulted from the passage of the powerful pressure wave through the star. In the hours following the neutrino burst, observers in Chile and in New Zealand found a new star in the LMC. The star grew 10 000 times brighter than its earlier incarnation as Sanduleak -69 202, and became the brightest star in the entire galaxy. For several months in early 1987, SN 1987A appeared as bright as the stars in the Big Dipper (or Plough!) even though the LMC is 1000 times as distant (see Figure 2). Astronomers measured the energy output of the star, which gave strong hints that radioactivity from elements made in the explosion provided the energy source to power the later stages of the supernova.

## NUCLEAR COOKING?

The radioactive material was predicted by models for supernova explosions. The idea is that oxygen or silicon outside the collapsing core could be briefly heated by the passage of the strong shock wave through the star. This would create conditions for cooking new elements to iron and beyond. Some of the nuclei formed in this way are not quite stable. For example, the models predict that this type of fusion would create the element nickel, in the form called $^{56}$Ni which has 28 neutrons and 28 protons. By radioactive decay, the nickel nucleus can emit a high energy photon (a gamma ray), and eject a positively charged electron and a ghostly neutrino. This nuclear prestidigitation changes a proton into a neutron and leaves the element cobalt in the form of a nucleus with 27 protons and 29 neutrons. But the cobalt nucleus isn't stable either: a lump of $^{56}$Co decays with a timescale of 77 days into the stable element iron in the form of nuclei with 26 protons and 30 neutrons. Each of these nuclear changes slides down the path toward the most tightly bound nucleus, and releases energy like a ball bouncing down a flight of stairs. In the case of SN 1987A, about 0.07 solar masses of radioactive nickel are produced. The amount sounds more impressive if you think of it as 20 000 Earth masses of seething radioactive waste. That radioactive source works like a battery to keep the supernova shining long after the initial blast has passed.

Although the shape of the light curve measured with optical telescopes is a strong hint that radioactivity lies beneath the appearances, for SN 1987A the real test comes from a new way of studying supernovae. Observations in the infrared, made from a high-flying airborne observatory, allow observations of material deep inside the expanding material of the ruined star. By flying the telescope above the absorbing water layers of the Earth's atmosphere, these observations reveal emission from cobalt atoms, part way down the chain of decay powering the supernova. The evidence suggests that the amount of cobalt decreases with time, just as expected if the cobalt is seeping away, changing into iron. Another product of the cobalt decay is the gamma ray emitted as the nucleus changes to iron. Again, new technology is needed. Balloons filled with helium lift gamma ray detectors 20 miles up to the edge of space. At these altitudes, the gamma rays from the supernova were measured in just the right amount to account for the observed energy output of the supernova. These new observations of SN 1987A have moved the idea of nuclear cooking in supernova explosions from a well-founded theory to an observed fact. This gives us confidence that other consequences of the theory – including the synthesis of the rare elements in the universe – are also correct, although they have not been subjected to direct tests.

## THE ENRICHED GET RICHER

Each supernova explosion yields the products of stellar evolution – helium and carbon, oxygen and silicon – into the gas between the stars. It also provides a pinch of the elements elaborated from iron seeds: palladium and silver, copper and tin, osmium and thorium. In time, the debris of the explosion mixes with the gas between the stars, and the specific result of each individual supernova becomes part of the general enrichment of the universe. In the history of our own Milky Way Galaxy, we have clear evidence that this process has been at work. By analysing the light from the atmospheres of stars, astronomers can deduce their chemical abundances. In some cases, we know the ages of stars, especially when stars are found in large spherical groups called globular clusters. Putting these two clues together, we find that there has been a profound change in the chemical composition of our galaxy over the last 10 thousand million years. The oldest stars, roughly 15 thousand million years old, have only about 1/1000 the abundance of heavy elements found in more recently formed stars. A star like the Sun, which is about 5 thousand million years old, has inherited antique gold from the generations that lived and died violently before its birth. The material of the Earth was formed in the cores of those ancient stars. Each of us is stardust.

## CATASTROPHE OR DISASTER?

A series of catastrophes has brought each of us to his current state. The sudden collapse of ancient stars has led to the aluminium and steel of modern telescopes that help us gather the clues that our hydrocarbon brains assemble into understanding. But I can't help thinking that a somewhat more sinister and threatening form of catastrophe is lurking in the back of those assemblies of ancient atoms. What about the danger from exploding stars?

The Sun is not likely to become a supernova. The core collapse that leads to explosions is restricted to more massive stars. Stars of the Sun's size can explode, but only when they are members of a binary system. The forecast for the Sun is continued sunny for the next 5 thousand million years or so. When the Sun has used significant amounts of the hydrogen in its core, it will change, eventually swelling to become a red giant star. This slow and gradual evolution will have some side effects. Here on Earth, the oceans will boil, temporarily reducing the price of fish and chips, eventually leading to the end of life on Earth. That's a disaster, but it doesn't have the sudden quality of a catastrophe.

What about the possibility of a nearby star exploding? Taking into account the Sun's orbit through the Galaxy and the current rate of supernova explosions, I calculate that the nearest explosion would be roughly 30 light years away. At that distance (5000 times closer than SN 1987A), a supernova explosion would be an awesome experience, shining a hundred times as brightly as the full moon. While this would spoil astronomical observations of other objects, it is hard to imagine wanting to work on anything else.

Still, even such a bright supernova would be much less than 1% of the Sun's brightness, and would only shine brightly for a few weeks. It seems unlikely to be very destructive, although it might disrupt some delicately poised systems such as the world's weather patterns. So except for a few hurricanes, droughts, and floods, the effects would be small – at least compared to melting the Earth.

There is one aspect of nearby supernovae that might have a significant effect. The magnetic spinning neutron star inside a supernova accelerates individual protons and electrons to very high energies, comparable to the energies in particle accelerators. Alternatively, as the expanding debris collides with the surrounding gas, its energy can accelerate particles to very high energies. We know that these processes are at work in supernova remnants, where powerful radio emission is produced when the fast-moving particles interact with the magnetic field in the remnant.

Our Galaxy, and the neighbourhood of the Earth, is awash with

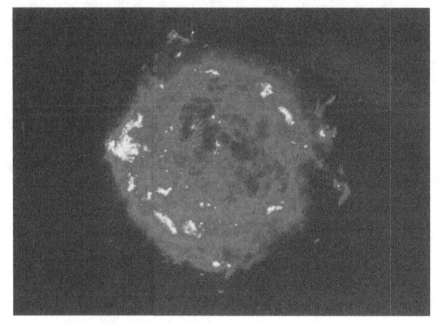

*Figure 9* The supernova remnant Cassiopeia A. The remnant of a supernova that exploded a few hundred years ago in our own galaxy. This image shows the radio emission from high energy particles interacting with magnetic fields in the interior of the remnant. The expanding remnant may reach a diameter of over 100 light years. It seems likely that the Earth has been inside supernova remnants in the past. The increased cosmic ray background may have increased the pace of mutation during those epochs.

high-energy particles of this general type. They have the quaint name 'cosmic rays' and the cosmic rays were what Zwicky emphasised in his brief 1934 prediction. The proton-decay experiments that detected neutrinos are deep underground to cut down on the background caused by these energetic and penetrating particles. Cosmic rays can damage genetic material, revising the genetic code without knowing the language of life. This cosmic cryptography leads to biological mutations. As individuals it is prudent to avoid having our genomes scrambled, but it is widely believed that natural selection provides the mechanism to sort through mutations and it can lead to original and perhaps improved species.

As the shock from a supernova explosion grows, the interior would be a region where the high-energy particle background would go up, not by 1%, but by a very large factor. The volume of supernova remnants is large enough that episodes of this type are likely to have occurred in the history of life on Earth. An expanding shell from a supernova would grow and, in the course of 10 000 years eventually envelop the Earth. The net result would be a few thousand years of increased rates of mutation – hastening the evolution of new species. Whether this is a disaster depends on whether you are the new species or an old one.

## FURTHER READING

Ferris, Timothy, *Coming of Age in the Milky Way*, New York: Morrow, 1988.
Goldsmith, Donald, *Supernova! The exploding star of 1987*, New York: St Martin's Press, 1989.
Kirshner, Robert P., 'Supernova – death of a star', *National Geographic*, May 1988.
Marschall, Laurence A., *The Supernova Story*, New York: Plenum, 1988.
Murdin, Paul, *End in Fire*, Cambridge: Cambridge University Press, 1990.

## 2

# The extinction of the dinosaurs

*WALTER ALVAREZ and FRANK ASARO*

Sixty-five million years ago, half of the life on Earth died out. The dinosaurs were wiped out, and their estate was inherited by the mammals – humble survivors of the event. As the descendants of those survivors, we cannot avoid the questions: Why? And what lessons does it teach us about life, the earth, and the cosmos?

For the last dozen years, scientists from around the world and in disciplines ranging from paleontology to astrophysics have mustered their observational skill, experimental ingenuity, and theoretical imagination in an effort to solve this mystery. Those of us involved in it have lived through long months of painstaking measurement, periods of bewilderment, flashes of insight, and times of excitement as parts of the puzzle fell into place.

How does one solve a puzzle like this mass extinction 65 million years ago? It's a little like reading a murder mystery novel, trying to recognise and interpret all the clues. But it's more like real detective work, because we're not limited to the hints the author provides; we are free to think up new approaches and to search for new kinds of evidence. The passage of so much time has also destroyed or degraded much of the evidence we are seeking, leaving only the most subtle clues to be found and interpreted, and so we must be content with circumstantial evidence and arguments about plausibility.

## THE START OF A MYSTERY

The sedimentary rocks that carry the record of life on Earth show a great discontinuity 65 million years ago. Animals like the dinosaurs and the ammonites, which had been abundant for tens of millions of years, suddenly disappeared forever, and many other groups of animals and plants were decimated before recovering. Geologists use this mass extinction, this punctuation in life history, to separate the Cretaceous period of Earth history from the subsequent Tertiary period, and refer to it as 'the K–T boundary' (Figure 1).

We first ask how long it took for this mass extinction to occur. Was it sudden – a few years or centuries – or was it a gradual event requiring millions of years to complete? Earlier geologists and paleontologists, influenced by the traditional preference in the Earth Sciences for slow, non-catastrophic explanations, assumed a gradual extinction, and at first the fossil record seemed to bear this out. However, when paleontologists looked at the single-celled marine animals called foraminifera, whose shells are extremely abundant, or at microscopic fossil pollen from plants, they found the extinction to be very abrupt. There seems to be a general relation: the smaller the organism, the more abundant its fossils are, and the sharper the extinction appears. Dinosaur fossils are rare, so it is difficult to be sure whether their extinction was sudden or gradual. Microscopic fossils are very abundant, and we now know that for those that died out, the extinction was extremely sudden.

With medium-sized fossils like those of the marine invertebrates, paleontologists are finding that the more closely they study the rock record, the sharper the extinction appears. A good example is the case of the ammonites – relatives of the modern chambered nautilus – a group that died out at the end of the Cretaceous. The best stratigraphic record of the ammonite extinction is found in coastal outcrops where the Spanish–French border reaches the Atlantic Ocean. Detailed studies of the Spanish outcrops at Zumaya, by Peter Ward of the University of Washington and his colleagues in 1986, suggested that the ammonites died out gradually, with one species disappearing

*Figure 1* The mass extinction of reptiles at the end of the Cretaceous (the Cretaceous–Tertiary,

after another through an interval of 170 metres, representing about 5 million years. However, in 1988, Ward studied two nearby sections in France, and discovered that all these ammonite species actually lived right up to the K–T boundary, where they suddenly became extinct. The apparently gradual extinction at Zumaya was due to local lack of

or K–T boundary) seems to have resulted from a huge asteroid impact.

fossil preservation; with the discovery of a locality where preservation was better, the abruptness of the extinction was undeniable (Figure 2).

So for fossils with a good stratigraphic record, the groups that died out did so abruptly, and it is reasonable to assume this was also the

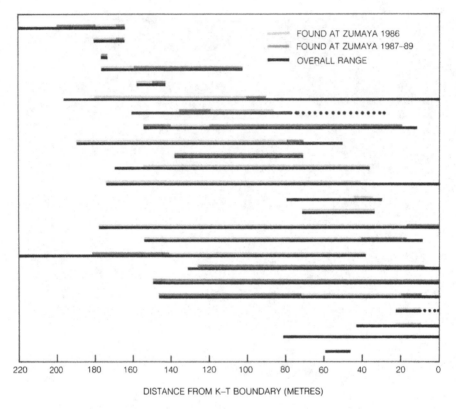

FOUND AT ZUMAYA 1986
FOUND AT ZUMAYA 1987–89
OVERALL RANGE

DISTANCE FROM K–T BOUNDARY (METRES)

*Figure 2* The sharpness of ammonite extinction at the K–T boundary increases as more data become available.

case for poorly recorded groups like dinosaurs that became extinct at about the same time. But how long, in years, did this abrupt extinction take? This is a difficult question, because of the lack of precision in the rock record. The basic time unit used by geologists is not one year, but one 'million-years', written '1 Myr'. It is a major accomplishment to date a rock with an accuracy of 1 Myr. *Intervals* of time between specific levels in the stratigraphic record can be determined somewhat better, in favourable circumstances, but it is hard to be more precise than 0.01 Myr.

In the deep-water limestones at Gubbio, in the Italian Apennines, Isabella Premoli Silva of the University of Milan, who discovered the K–T boundary, found a 1cm layer of clay separating the Cretaceous and Tertiary limestones. In 1976, one of us (Alvarez), working with Premoli Silva, William Lowrie of ETH-Zürich, Giovanni Napoleone of the University of Florence, and Alfred G. Fischer, Michael A. Arthur and William M. Roggenthen of Princeton University, showed that this clay layer fell within a 6m thickness of limestone deposited during the 0.5 Myr interval of reversed geomagnetic polarity designated '29R'. On the face of it, this suggested that deposition of the clay layer, and thus the mass extinction, lasted no more than about 0.001 Myr, so brief that we can switch from the geologic to the human scale, and call it 1000 years. Jan Smit of the University of Amsterdam did the same analysis at Caravaca, in southern Spain, where the stratigraphic record is more precise, and estimated that the extinction lasted no more than 50 years. By geologic standards, this is blindingly fast!

## IMPACT ON THE EARTH

In the late 1970s, working with our Berkeley colleagues Luis W. Alvarez and Helen V. Michel, we thought we could make a small contribution to solving the dinosaur-extinction question with a new method for determining more accurately how long the Gubbio clay layer took to be deposited and thus, indirectly, how long the extinction event lasted. Unfortunately, our method didn't work, but in the process we stumbled onto the first real clue. That's what a scientist needs: a lot of hard work and an occasional lucky break.

We knew that iridium and the other platinum-group metals are rare throughout the solar system (iridium occurs with an abundance of about 500 parts per billion in the primitive stony meteorites called carbonaceous chondrites, for example). But they are much rarer still in the rocks of the Earth's crust, (iridium abundance is about 0.01 parts per billion in the deep-water limestones at Gubbio). Presumably this rarity of the platinum-group elements in the Earth's crust occurs because most of the Earth's allotment of these elements, which alloy with iron, was absorbed into the iron core when it formed, leaving them depleted in the crust. We suspected that such a rare and insoluble element as iridium would enter deep-sea sediments predominantly through the continual rain of micrometeorites, sometimes called 'cosmic dust'. What's more, iridium could be measured at very low concentrations using neutron-activation analysis, a technique in which neutron bombardment converts the iridium into a radioactive and hence detectable form.

Our idea is shown in Figure 3. Using the assumption that most of the iridium in the clay arrived as cosmic dust, we planned to use iridium as a sedimentation-rate meter to distinguish between several possible origins of the Gubbio clay bed. The limestone contains about 5% clay and the 1 cm thick K–T boundary bed contains about 50% clay. To simplify the problem we focused on the iridium:clay ratio by removing the limestone component before analysis. The sedimentation rate of the clay should then be inversely proportional to the iridium abundance. One set of possibilities had the limestone deposition stopping or slowing down for some 10 000 years. If the clay deposition continued at the same rate, the abundance of iridium in the K–T boundary clay layer would be unchanged from its value in the clay above and below. If the clay deposition rate slowed down, the iridium abundance would be higher, and if it speeded up the iridium abundance would be lower. A very similar technique, using iridium from cosmic dust for determining the ages of soils, had already been proposed and tested by Andrei Sarna-Wojcicki of the US Geological Survey.

In June of 1978, our first Gubbio iridium analyses were ready, and we scanned them excitedly, to find out whether the extinction event had lasted 10 000 years, or only a few years. Imagine our astonishment and confusion when we saw that the Gubbio boundary clay and the immediately adjacent limestone was much richer in iridium than we could explain by any of our scenarios. This feature came to be called 'the K–T boundary iridium anomaly'. The amount of iridium in the anomaly was comparable to the amount of iridium in all the rest of the 500 000 years of magnetic polarity interval 29R put together. This accidental discovery was our lucky break.

Where could so much iridium have come from? Clearly it could not come from the usual infall of micrometeorites. Evidently, we were

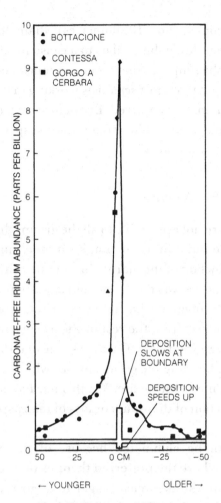

*Figure 3* Using the iridium : clay ratio in Gubbio limestone to determine sedimentation rates. The very high levels of iridium at the K–T boundary were probably due to the catastrophic impact of a 10 km diameter asteroid. Other extinction scenarios in which the rate of limestone deposition varied (bottom) cannot explain so marked an anomaly.

seeing a sudden and abnormally large dumping of iridium onto the Earth's surface. For a whole year we debated what could have done this, testing and rejecting idea after idea. Finally, in 1979, we proposed the one solution which had survived all this testing. The excess iridium in the boundary clay, we concluded, was due to the impact on the Earth of a large asteroid or comet, about 10 kilometres in diameter.

The unexpected clue from the iridium had led us to a cause for the dinosaur extinction: catastrophic impact. Since then, so much confirming evidence has come to light that most scientists working in this field are persuaded that a great impact occurred. But before we look at these developments, let us pause and think about other suspects. What else might have killed the dinosaurs?

## OTHER POSSIBLE CAUSES

Some of the suspected causes are not applicable to all the groups that became extinct. The venerable notion that mammals ate the dinosaurs' eggs cannot explain extinction of the marine foraminifera and ammonites. A more interesting case was the ingenious suggestion by Stefan Gartner of Texas A & M University that in the Late Cretaceous the Arctic Ocean had been isolated from the rest of the oceans and filled with fresh water by converging rivers. Gartner suggested that this huge lake had overflowed, flooding the ocean surface with fresh water, which would have poisoned marine life. Yet this mechanism could not account for the extinction of dinosaurs or loss of many species of land plants.

Other suspected causes can be eliminated because of timing. Changes in climate and sea level are the preferred theories of a few scientists. However, these changes take much longer to occur than did the extinction, they do not seem to have coincided with the extinction, and they have occurred repeatedly in the Earth's history with no accompanying mass extinction. They do not seem like promising possibilities.

What about volcanism, which a few other detectives consider a prime suspect? The strongest evidence implicating volcanoes is the existence of an enormous outpouring of basaltic lava in India – the Deccan Traps – at approximately the time of the mass extinction 65 Myr ago. Recent paleomagnetic work by Vincent Courtillot and his colleagues at Paris confirms studies by many workers, going back to the 1950s, which have shown that most of the Deccan Traps lava erupted during a single episode of reversed polarity, overlapping slightly into the preceding and subsequent normal polarity intervals. New age dates by the Paris team indicate that the Deccan reversed polarity interval was probably the same as reversed polarity interval 29R, during which the K–T extinction occurred, although it might match with the reversed interval either before or after 29R. Reversed polarity intervals near the K–T boundary lasted about 0.5–1.0 Myr, and since the Deccan Traps began and ended during normal times, they must represent a minimum of 0.5 Myr.

Until recently, most scientists interested in the problem have not considered volcanism a serious suspect. Even if the timing was approximately right, the 0.5 Myr duration of Deccan volcanism was too long to account for a mass extinction that took place in 0.001 Myr or less. And most important, impact coherently explains the five lines of critical evidence we are about to consider, whereas volcanism does not. Yet the coincidence in age of the extinction and the Deccan volcanism is striking, and we should not ignore coincidences. Could volcanism have been triggered by impact? We will return to this question at the end.

## EVIDENCE FOR IMPACT

What, then, is the evidence for major impact? Since 1979, anomalously high levels of iridium have been found at the K–T boundary at many localities around the world (see Figure 4). Our group, including our Berkeley colleague Alessandro Montanari and many other collaborators, has found the anomaly in Italy, Denmark, New Zealand, Haiti, Canada, Spain, many sites in the US, and sediment cores from the Pacific, Atlantic, and Indian Oceans, and the Weddell Sea off Antarctica. Jan Smit of the University of Amsterdam has found it in Spain. R. Ganapathy of the J. T. Baker Chemical Company and his colleagues found it in Texas. Kenneth J. Hsü of ETH-Zürich and his collaborators have found it in the South Atlantic. Frank Kyte, Zhiming Zhou and John Wasson of UCLA have found it in the Pacific Ocean. Charles J. Orth and co-workers at Los Alamos and Charles

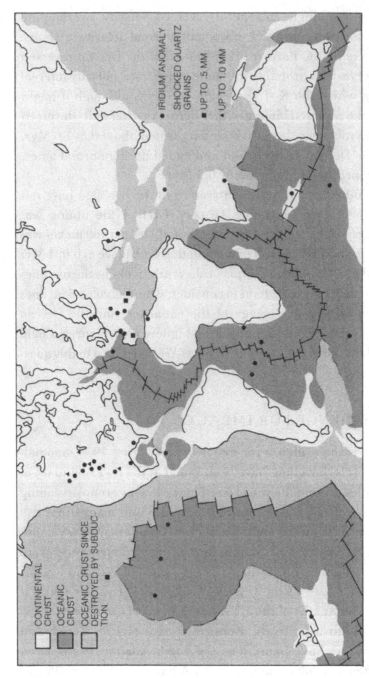

*Figure 4* World distribution of iridium anomalies and impact-generated shocked quartz crystals at the K–T boundary. The continents are shown as they were 65 million years ago.

Pillmore, Robert Tschudy, Glen Izett, and Hugh T. Millard of the US Geological Survey have found it in many non-marine boundary sediments in New Mexico and Colorado. In fact, more than 100 scientists in 21 laboratories in 13 countries have found the K–T iridium anomaly at over 100 sites around the world. And by now we have analysed enough other stratigraphic levels to know that iridium anomalies are very rare; as far as we know, the K–T anomaly is unique.

The iridium anomaly is well explained by impact. But in 1983, another possibility arose when William Zoller and his colleagues at the University of Maryland discovered high concentrations of iridium in aerosols erupted by Kilauea Volcano and collected on filters 50 km away. However, there are several other elements, including gold, that are depleted in the Earth's crust relative to their meteorite abundance; they all show anomalous concentrations at the K–T boundary, although the others are more difficult to measure than iridium and therefore less often studied. In the Kilauea aerosols, the abundance ratio of gold to iridium was at least 35 times higher than in the carefully studied K–T boundary at Stevns Klint, in Denmark. In 1986, I. Olmez and his associates in the Maryland group found the same high gold:iridium ratio in samples taken directly from the volcanic vent at Kilauea, and still more detailed work at Kilauea by Bruce Crowe of Los Alamos National Laboratory and his associates showed the Kilauea gold:iridium ratio to be 19–400 times higher than in the K–T boundary at Stevns Klint. In contrast, if the Stevns Klint gold:iridium ratio as reported by Miriam Kastner of the Scripps Institute of Oceanography and the Berkeley group is compared with the ratio in the most primitive meteorites (Type I carbonaceous chondrites), the ratios agree within 5 per cent. Similarly, Carl Orth and his associates reported a gold:iridium ratio at a non-marine K–T boundary site at Berwind Canyon in Colorado that agreed with meteorite values. So it is very reasonable for the K–T iridium to have come from an impact, but highly unlikely that it came from a volcano like Kilauea.

Further developing the idea in the previous paragraph, we can test whether meteorite debris is present in rocks by checking the abundance ratios of all the platinum-group elements. These elements (ruthenium, rhodium, palladium, osmium, iridium and platinum) have similar chemical behaviour; they are somewhat resistant to chemical alteration and they are depleted in the crust of the Earth compared to meteorites. The ratios in meteorites are distinctly different from those in the crust. R. Ganapathy made the first measurement of platinum-group element ratios in the K–T boundary from Stevns Klint in 1980, followed by John Wasson and his associates at UCLA shortly afterwards. Both groups concluded that the platinum-group elements in the K–T boundary had an extraterrestrial origin. Vladilen S. Letokhov and George Bekov from the Institute of Spectroscopy in Moscow concluded from rhodium:iridium ratios measured by laser techniques that a K–T boundary in Turkmenia, in the Soviet Union, had an extraterrestrial origin. Bekov and Asaro have found that the ratios of abundances of ruthenium and rhodium measured in Moscow and iridium measured at Berkeley can distinguish stony meteorites from terrestrial and mantle-derived samples. We conclude that the chemical evidence supporting an extraterrestrial source for the K–T iridium is very strong.

Not only element ratios, but also isotopic ratios have been used to test the impact hypothesis. Karl Turekian at Yale pointed out that the element rhenium, which is much more abundant on continents than in meteorites or the Earth's mantle, has an isotope, $^{187}Re$, which decays radioactively to the osmium isotope, $^{187}Os$. As a result, osmium in continental rocks has a higher ratio of the daughter isotope, $^{187}Os$, to ordinary osmium, $^{186}Os$, than one finds in asteroids, comets, or the Earth's mantle, which are not enriched in the parent isotope, $^{187}Re$. The ratio $^{187}Os : ^{186}Os$ thus provides a fingerprint for distinguishing osmium derived from continents. Measuring K–T boundary samples from Denmark and New Mexico, Jean-Marc Luck and Turekian found osmium isotopic ratios which suggested that the predominant part of the osmium came not from a continental source, but probably from an extraterrestrial object, although they could not completely rule out a mantle origin.

In 1981, Jan Smit discovered mineral spherules up to about 1 mm in diameter in the Caravaca K–T clay, and Alessandro Montanari confirmed their presence in the Italian boundary as well. Studies of the chemistry, mineralogy, and texture of these spherules by Smit, Montanari, and their colleagues show convincingly that the spherules originated as droplets of the target rock, shock-melted by impact, and rapidly cooled during ballistic free fall outside the atmosphere before they re-entered and were deposited and chemically altered in the boundary clay. The spherules are thus the crystallised equivalent of the glassy tektites and microtektites that are the known result of smaller, younger impacts.

Perhaps the most convincing of all the evidence for impact at the K–T boundary was the discovery of shocked grains of quartz by Bruce Bohor of the US Geological Survey and Donald M. Triplehorn of the University of Alaska and their detailed study by E. E. Foord, P. J. Modreski, and Glen Izett, also of the US Geological Survey. These quartz grains carry the multiple, intersecting planar 'lamellae', or bands of deformation that are diagnostic of hypervelocity shock. Quartz grains with these lamellae are found only in known impact craters, at the sites of nuclear tests, in materials subjected to extreme shock in the laboratory, and in the K–T boundary.

Recently, there has been a technical debate as to whether explosive volcanism might also produce shocked quartz. It now seems agreed that the weaker shocks of volcanic explosions can produce some deformation in quartz, but that the distinctive multiple lamellae seen in the K–T boundary quartz grains form in nature only in the more extreme shocks generated by impact.

Still other evidence ties the K–T shocked quartz to impact. John McHone of Arizona State University has identified stishovite, the extreme high pressure polymorph of quartz, in the shocked grains. Mark H. Anders of Columbia University and Michael R. Owen of St Lawrence University have used cathode luminescence to show that the K–T shocked quartz comes from a plutonic and sedimentary source, not a volcanic source. Finally, an impact crater has been located beneath the glacial drift at Manson, Iowa, with a quartz-rich bedrock, and in a location suitable to explain the size and abundance distribution of the boundary quartz grains. Detailed studies of the Manson crater by Jack Hartung of the Lunar and Planetary Institute, Raymond Anderson of the Iowa Geological Survey, and Michael Kunk and colleagues of the US Geological Survey, show it to have an age indistinguishable from the age of the K–T boundary.

Not only the nature of the unusual features of the K–T boundary but also their distribution strengthens the case for impact. Since we discovered the iridium anomaly in Italy, anomalous iridium has been found at over 100 K–T boundary sites around the world. The characteristic spherules of the K–T boundary have also been found in more than 60 sites and they are apparently also worldwide. Shocked quartz has been found at 26 widely scattered sites, and it may also be distributed world wide, although it seems to be more abundant and in larger grains close to its apparent site of origin at the Manson Crater in Iowa. The apparent worldwide distribution of these features is in itself evidence for impact rather than volcanism. A volcanic eruption takes place at the base of the atmosphere, and eruption products are injected into the lower atmosphere. Fine aerosols may be carried long distances in the atmosphere, although because of circulation patterns they do not easily cross the equator. However, larger ejecta grains are quickly slowed down by atmospheric drag and they fall out. The spherules and shocked quartz are the size of grains of sand, and they simply could not be carried around the world within the atmosphere.

This problem is removed in the case of large-body impact, for the particles are carried outside the atmosphere and travel on ballistic trajectories until they re-enter the atmosphere at points all around the globe, and then fall to the ground. There are two features of large-body impact that result in ejecta leaving the atmosphere. In the first place, the incoming body had an estimated diameter of 10 km, which is close to the half-height of the atmosphere – the level below which half of the air is confined. At a velocity of more than 11 km/sec, an object this size simply rams a huge hole in the atmosphere. The explosion, occurring when the kinetic energy of the object is released upon impact, takes place at the base of this hole and much of the ejecta leaves the atmosphere through this escape route before the hole has time to close up.

The second mechanism for getting ejecta outside the atmosphere requires some understanding of very large explosions on the Earth's surface. The largest explosions that human beings have produced result from nuclear weapons with energy releases of the order of one megaton. When this much energy is suddenly released in a violent explosion it creates a fireball of incandescent gas which expands until it reaches the same pressure as the surrounding atmosphere, and then rises until it reaches a level in the atmosphere at which its density is the same as that of the surrounding air. At this point, usually at around ten kilometres elevation, the gas spreads laterally, forming the head of a mushroom cloud. Human beings have not produced explosions with energies of a thousand megatons, but explosions this size have been modelled by computer. The result is a fireball of such enormous proportion that it never reaches pressure equilibrium with the atmosphere. In fact, as this gigantic expanding fireball rises to levels where the density of the atmosphere is significantly declining, the rise of the expanding fireball accelerates and the gas leaves the top of the atmosphere with velocities fast enough to produce escape from the Earth's gravitational field.

In the case of the explosion produced by impact of a comet or asteroid about ten kilometres in diameter (the size estimated for the K–T boundary event), we are not dealing with one megaton, or one thousand megatons, but one hundred million ($10^8$) megatons. It is important to remember that this energy does not come from a nuclear explosion; it is simply the kinetic energy released by a large object striking the Earth at high velocity. When we consider that the world nuclear arsenal at the present time consists of roughly 10 000 warheads of one megaton energy each, we see that the K–T boundary event released non-nuclear energy equivalent to 10 000 times the energy of all of the existing nuclear weapons. The fireball from such an enormous explosion simply bursts out of the top of the atmosphere, carrying any entrained ejecta along with it and sending this material into ballistic orbits which easily carry it to any place on Earth. This mechanism provides a good explanation for the worldwide distribution of iridium, spherules, and shocked quartz in the K–T boundary layer.

For a scientific detective, the evidence from the K–T boundary clay that we have just discussed (iridium anomalies, platinum-group element ratios, osmium isotopic ratios, spherules, shocked quartz, and worldwide distribution of at least some of these features) shows us that a potential killer, in the form of giant impact, was on the scene at the time of the extinctions. But we also have to establish means – that is, what were the specific mechanisms by which the mass extinction was produced?

## KILLING MECHANISMS

In the 1980 paper in which we proposed an impact at the K–T boundary, we discussed one killing mechanism that would result from giant impact. This is the darkness that would result from the spreading of dust-sized impact ejecta into the atmosphere worldwide. It is important to realise that ejecta dust will be formed even if the impact is in the ocean, because an object ten kilometres in diameter is falling into an ocean typically only five kilometres deep, so it is like throwing a large rock in a shallow puddle. Our original rough estimates of the resulting darkness were carefully refined in computer simulations by Richard Turco of R & D Associates, Brian Toon of NASA Ames, and their colleagues, who have shown that following impact of a ten-kilometre object, the lofted dust would make it so dark around the world that, for a period of a few months, you literally could not see your hand in front of your face. This lack of sunlight would halt photosynthesis, causing collapse of food chains worldwide, and it would also produce extremely cold temperatures. This scenario has come to be called 'impact winter'. Immediately after their work on impact winter, Turco, Toon, and their colleagues went on to consider the rather similar 'nuclear winter' which would follow a full-scale nuclear war.

In 1981, Cesare Emiliani of the University of Miami, Eric Krause of the University of Colorado, and Eugene Shoemaker of the US Geological Survey pointed out that if the K–T impact occurred in the ocean, it would not only eject solid rock dust, but would also disperse a great deal of water vapour into the upper atmosphere worldwide.

The dust would settle to the Earth in a few months, but the water vapour would remain aloft for much longer. Water vapour traps solar heat, and so a phase of greenhouse heating would follow the cold of the impact winter. More recently, John D. O'Keefe and Thomas Ahrens of Caltech have suggested that if impact occurred in an area of limestones, it would release a large amount of $CO_2$, another greenhouse gas, which could result in an episode of very high temperatures. It thus appears that major impact would lead to a variety of severe climatic disturbances, which might well be lethal to various groups of plants and animals.

In addition to heating and cooling, the killing mechanisms that would result from a large impact could include acid rain. John Lewis, G. Hampton Watkins, Hyman Hartman, and Ronald Prinn of MIT, calculated that shock heating of the atmosphere during impact would overcome the activation-energy barrier that normally keeps the nitrogen and oxygen in the air from combining. The resulting NO would go through a series of reactions involving the $NO_x$ compounds, eventually raining out of the air as nitric acid. This truly serious acid rain may well explain the widespread extinctions of marine invertebrate plants and animals whose calcium carbonate shells are soluble in acidic water.

Another killing mechanism came to light when Wendy Wolbach, Ian Gilmore, and Edward Anders of the University of Chicago discovered large amounts of soot in the K–T boundary clay layer. They pointed out that if the boundary layer represents only one year or a few years, the amount of soot present represents burning of vegetation equivalent to about half of the present day forests. The fires could have been ignited by infrared radiation from the incandescent front of the incoming object throughout the area where the actual impact was visible. Recently Jay Melosh of the University of Arizona and his colleagues have suggested that ejecta re-entering the atmosphere throughout the rest of the world would provide a heat source sufficient to ignite forest fires worldwide.

Detailed studies of the K–T boundary sediments may eventually be able to provide evidence for choosing among the suggested killing mechanisms. For example, Alessandro Montanari has shown, from dissolution patterns in the limestone, that the ocean-bottom waters in Italy were acidic immediately after the K–T extinction. In rock-magnetic and geochemical work with William Lowrie, we have shown that the normally oxidising character of the Italian bottom waters was briefly changed to a reducing condition immediately after the K–T event, which may indicate a massive kill of the oceanic biota.

Edward Anders has summarised current thinking on the range of killing mechanisms that would result from impact. In view of this Inferno of gruesome environmental disturbances, it does not seem at all surprising that large-body impact could produce a devastating mass extinction. In fact, Norman Sleep of Stanford University and his colleagues Kevin Zahnle, James Kasting, and Harold Morowitz have suggested that in the very early history of the Earth, when objects substantially larger than the K–T boundary impactor struck the Earth frequently during its phase of accretion, incipient life may have been extinguished by impact more than once before conditions became sufficiently calm for life to be an enduring phenomenon.

## HINTS OF A DEEPER MYSTERY

Where do we stand now in our investigation of this mystery? We have established that there were victims of extinction, and that they were killed very rapidly by geological standards. We have a cause – large-body impact – whose existence at the time of the event is definitely established. Have we solved the problem and understood what happened? Until recently we thought so, and some people continue to think that this is all there is to the story. Perhaps this *is* the whole story, but Nature, like a good mystery writer, seems to be leaving us clues that there is more to the story. The role of impact seems firmly established but there are intriguing indications that the event may have involved more than a single, isolated impact.

The first suggestion of a more complicated extinction event came in 1984 when David Raup and Jack Sepkoski of the University of Chicago published an analysis of the age ranges of families of fossil organisms, which seemed to indicate that mass extinctions have occurred in the Earth's past at regular, periodic intervals of 32 million years. Their work offered quantitative support for the 1977 suggestion by Fischer and Arthur that various aspects of life and climate are cyclical with a period of about 32 million years. This was an astonishing result. We had strong evidence that at least one of those mass extinctions (at the K–T boundary) had been caused by an impact, and it was hard to envision a process more random in time than large body impact on the Earth.

Like most other scientists working on the K–T extinction, we were very sceptical of the Raup and Sepkoski results. Our colleague at Berkeley, astrophysicist Richard Muller, was also sceptical, but he took the trouble to reanalyse the data of Raup and Sepkoski, and convinced himself that their periodicity was real. Seeking a mechanism for producing mass extinctions caused by impact and occurring periodically in time, Muller teamed up with Marc Davis of Berkeley and Piet Hut of the Institute for Advanced Studies in Princeton. They realised that if the sun had a dim, unrecognised companion star orbiting the sun every 32 million years, the companion star, on its closest approach to the sun, could disturb the orbits of comets on the outer fringe of the solar system. The disturbance would send a storm of comets into the inner solar system, which would last for one or two million years, until the inner solar system was cleaned of comets by impact or by their ejection from the inner solar system. Davis, Hut, and Muller suggested that if such a dim companion star were found, it should be named 'Nemesis'. Meanwhile, Daniel Whitmire of the University of Southwestern Louisiana and Albert Jackson of Computer Sciences Corporation independently proposed the same companion star hypothesis. The Nemesis theory thus provided a mechanism for creating periodic comet storms, during each of which there would be a greatly increased probability of one or more large impacts on the Earth capable of producing mass extinctions. In addition, two other ways of producing periodic comet storms, now thought to be less likely than the companion star hypothesis, were proposed – one based on vertical oscillations of the solar system through the plane of the galaxy proposed by Michael Rampino and

Richard Stothers of NASA Goddard Institute, and the other based on a hypothetical tenth planet in a peculiar orbit, suggested by Whitmire and John Matese, his colleague at the University of Southwestern Louisiana.

If the Nemesis hypothesis is correct, craters on Earth should show the same periodicity in their ages. Richard Muller and W. Alvarez then analysed the time sequence of terrestrial crater ages, and discovered that they showed essentially the same periodicity as the mass extinction. From that point on we have felt that the Nemesis hypothesis must be taken seriously.

One way to take the hypothesis seriously is to search for the companion star itself. It turns out to be very difficult to find a dim red star close to the sun, when one has no idea in which direction to look. Muller and Saul Perlmutter at Berkeley are now about halfway through a computerised telescopic search for a star with the characteristics of Nemesis; they should complete this search in a couple of years. Meanwhile, new analyses of the time sequence of extinctions and craters have raised questions about whether they are strictly periodic. Unfortunately, the small number of extinctions and craters and the poor quality of the evidence concerning their age make it difficult to be sure if these events are periodic or not. To some extent then, the questions of periodic extinction and the Nemesis hypothesis must remain on the back burner until there is new information to discuss.

However, those early indications of a more complex link between impact and extinction have been followed by newer lines of evidence. One such line of evidence emerges from the question 'where did the K–T boundary impactor hit?' The impact should have left a crater about 150 kilometres in diameter. It has always been a major disappointment that no one has found this crater. However, there are many places in the world where one could hide a crater this size, for example under the Antarctic ice sheet. In addition, 20 per cent of the surface of the Earth that existed at the time of the impact was ocean floor which has subsequently been consumed in subduction zones.

Yet, there is now a strong suggestion that at least two nearly simultaneous hits occurred. Manson Crater in Iowa, already mentioned, is a very strong candidate as the source for the shocked quartz. Its age is turning out to be indistinguishable from that of the K–T boundary, but at 32 kilometres in diameter, Manson Crater is far too small to be the only K–T boundary crater. In the list of more than 120 terrestrial impact craters compiled by Richard Grieve of the Geological Survey of Canada, there are several craters about 30 kilometres in diameter, and this list represents a very incomplete sample of the total population of terrestrial craters. It is clear from the list of known impact craters that craters with diameters of 30 kilometres form far more frequently than mass extinctions occur. If the Manson Crater is confirmed as being exactly the same age as the K–T, it must still be only part of the story; we are still looking for at least one much larger crater, probably located on oceanic crust.

Given the tiny size of the Earth as a target, how could multiple impacts occur in a short period of time? Eugene Shoemaker and Piet Hut have identified several possible ways of getting multiple impacts, with each mechanism having a characteristic time span over which the multiple impact events would occur. One suggested mechanism can be eliminated as ineffective: the object does not simply break up in the atmosphere, because the atmosphere is too thin. The diameter of the impacting object, about 10 kilometres, is about the same as the depth below which half of the atmosphere is confined.

However, two different mechanisms could give multiple impact on the same day. A number of craters on the Earth, for example Clearwater Lakes in Canada, and on the moon and other planets are double or multiple, suggesting that some asteroids may travel as two or more objects mutually orbiting each other. Alternatively, the Earth might be struck by two or more large fragments of a comet nucleus which was already breaking up and dispersing.

Multiple impacts over time spans of thousands or tens of thousands of years could occur if a fragmenting and dispersing comet nucleus left several large fragments in the related orbits of an Earth-crossing meteor stream. Finally, the comet storms that would be unleashed on the inner solar system by Nemesis or by an unrelated passing star would greatly increase the probability of multiple impacts during the million-year duration of the storm. None of the scenarios for multiple impact is as yet the solution of choice, but they indicate that there are more possibilities for complex impact crises than we once realised.

Finally, we should note a very intriguing discovery by Meixun Zhao and Jeffrey Bada of the University of California at San Diego. In chalk layers just above and just below the boundary in Denmark they found occurrences of amino acids of types which are not used by life on Earth and which therefore are essentially absent in the terrestrial environment, but which are known to occur in carbonaceous chondrite meteorites. It seems unlikely that amino acids could survive the heat of the fireball from a major impact, and in fact Zhao and Bada found that the K–T boundary clay contained none of the exotic amino acids; they occurred only in thin stratigraphic intervals just above and just below the boundary. How did these amino acids get there? It is hard to escape the conclusion that they are of extraterrestrial origin. Kevin Zahnle and David Grinspoon of NASA Ames Laboratory have proposed that during an extended interval, the Earth was receiving a great deal of dust from a disintegrating comet; the extraterrestrial amino acids would have accompanied that dust, but fireball heat would have destroyed amino acids contained in a large body that impacted during that interval. As with the evidence for periodicity and the suggestions of multiple impact, the extraterrestrial amino acids seem to be clues for a more complex impact crisis, and a warning that we have not yet completely solved this mystery.

## WAS VOLCANISM INVOLVED?

Let us now return to the question of volcanism, the alternate suspect favoured by a few scientific detectives. Could volcanism have played a role in the increasingly complicated extinction scenario that we are being forced to consider? Let us first look at the evidence found in the boundary clay. We have seen that all of the peculiar evidence from the boundary clay is coherently explained by impact. On the other hand, volcanism does not explain the evidence in the boundary clay. Iridium has indeed been found in the vapour phase escaping from

Kilauea, but in that gas, gold and iridium do not occur in the ratio characteristic of both the boundary clay and of carbonaceous chondrite meteorites. Volcanic explosions may produce some slight deformation in quartz grains of the surrounding rock, but it is now agreed that they do not produce the characteristic intersecting lamellae of the K–T boundary shocked quartz, nor could they produce the extreme high pressure polymorph of quartz, stishovite, which has now been found in the K–T quartz. Glassy spherules may be produced in some quiet volcanic eruptions, but quiet volcanism offers no mechanism for distributing these particles worldwide. In fact, if the nature of the boundary clay were the only evidence we had to consider, there would be absolutely no reason to think that volcanism played any part in the K–T boundary extinction.

The case for volcanism rests instead on a question of timing. As mentioned earlier, new age information on the Deccan Traps of India argues that these basaltic eruptions took place at roughly the same time as the K–T boundary extinction. Supporters of volcanism therefore say that the Deccan Traps and the K–T extinction coincided in time. However, as discussed above, magnetic polarity evidence indicates that the Deccan Traps were erupted over a period of at least 500 000 years, whereas the duration of the extinction event is probably less than 1000 or even less than 100 years. One cannot meaningfully say that events with such different time scales were synchronous any more than one can say that the 1943 World Series was synchronous with the Second World War.

And yet the age of the Deccan Traps *is* intriguing. The Deccan Traps represent the greatest episode of volcanism on land in the last 250 million years (although much greater outpouring of lava goes on continually at the mid-ocean ridges). And during that interval of volcanism, or perhaps slightly before or slightly after, there occurred the greatest mass extinction and the greatest impact that we know of in the last 250 million years. No-one can afford to ignore that kind of coincidence.

What could this coincidence mean? It seems possible that Deccan volcanism might have been triggered by large impact. A few minutes after impact, the initial crater would be 40 kilometres deep, and one suspects that the release of pressure might allow the hot rock of the underlying mantle to melt if the impact were in an area of high thermal gradient. Our Berkeley group suggested this explanation for the Deccan Traps back in 1982; it seemed more reasonable to think that impact could trigger volcanism than that volcanism could trigger impact! However, authorities on the origin of large-scale volcanic provinces, like our colleague at Berkeley, Mark Richards, find it technically very difficult to explain the onset of large-scale basaltic volcanism by impact.

So where does that leave us? The evidence overwhelmingly supports impact but seems to require a complicated impact scenario on which there is not yet any agreement. The evidence from the boundary clay does not argue for volcanism, and yet the Deccan volcanism is implicated by a timing coincidence too extraordinary to ignore. In the last few years, the impact-volcanism argument has tended to become polarised. Supporters of impact have tended to ignore the Deccan Traps as irrelevant, whereas supporters of volcanism have tried to explain away the evidence for impact by suggesting that it is also compatible with volcanism.

Our sense is that we are in a Hegelian situation, with an impact thesis and a volcanic antithesis in search of a synthesis whose outlines we cannot yet see with any clarity. And yet, although the mystery has not been completely solved, we have already learned a great deal, and we will close this progress report with a brief discussion of what the research on impact and mass extinction has already taught us.

## LESSONS

In the late eighteenth and early nineteenth centuries, when the study of the Earth was first becoming a science, there was a long battle between catastrophists and uniformitarians – between people who thought that sudden great events were very important in the evolution of our planet, and those who wished to attribute all of the history of our planet to slow gradual change (see the discussion of catastrophism by Martin Rudwick in this volume). The uniformitarians, particularly the English geologist Charles Lyell, so thoroughly won this battle that generations of geology students have been taught that any sort of catastrophism is nonscientific, and that the only true geology is strictly uniformitarian and gradualistic. Perhaps this view of Earth history reflected the wishful thinking of Victorian English gentlemen like Lyell as to desirable rates of change in human history. Nevertheless, in the century and a half since Lyell, human history has witnessed one violent turnover after another, and it is ironic that geologists should have maintained the uniformitarian faith through all those social disturbances. The Universe is a violent place, as astronomy has taught us (see Robert Kirshner on Supernovae in this volume) and we are now seeing that the history of the Earth has also had its violent episodes.

Because of their uniformitarian training, most geologists have long rejected the notion that large-body impacts could have played a substantial role in the history of the Earth. As Eugene Shoemaker, perhaps our foremost authority on planetary geology, has written, most geologists are reluctant to believe that 'stones the size of hills or small mountains can fall out of the sky'. Or as an earlier geologist, Thomas Jefferson, is supposed to have said when told that two Yale scientists had reported a meteorite fall, 'It is easier to believe that Yankee professors would lie than that stones would fall from heaven'. Perhaps this bias against impacts as a geological process should have died when geologists and astronauts went to the moon and showed beyond any reasonable doubt that impact was the dominant process in the formation of the Earth's surface. However, as Ursula Marvin of the Harvard–Smithsonian Center for Astrophysics has pointed out, it is a strange irony of science that in the late 1960s, as the Apollo lunar programme was going on, Earth scientists were swept up in a different breakthrough – the recognition of plate tectonics. Nothing could be more uniformitarian than plate tectonics, and despite the overwhelming evidence for impact emerging from the space programme, the uniformitarian bias of geology was only strengthened through the discovery of plate tectonics.

With the evidence that a giant impact was responsible for the extinctions at the end of the Cretaceous, the catastrophic viewpoint has finally become respectable. Even more dramatically, the origin of the moon is now widely attributed to impact on the early Earth of an object the size of Mars. Future geologists, with the intellectual freedom to think in both uniformitarian and catastrophic terms, have a better chance of really understanding the processes and history of our planet than did an earlier generation, shackled by an outdated uniformitarian viewpoint.

If indeed a chance impact 65 million years ago wiped out half of the life on Earth, then catastrophes of this type must play an important role in the evolution of life on our planet. Wallace and Darwin realised that evolution proceeds by the survival of the fittest, and there is no reason to think that they were wrong in this regard. However, the study of impacts and mass extinctions makes us realise that an additional factor must be involved in driving evolution. Successful species must not only be well adapted, they must also be lucky. Survival of the fittest cannot be the only factor; surely half of the species on Earth did not suddenly become unfit at one abrupt moment 65 million years ago. Perhaps we should say that superb adaptation fits a species to survive in normal times, but only good luck gets it through the occasional disaster.

If whole arrays of well adapted organisms are wiped out from time to time through the chance disaster of large impact, it means that the history of life is not foreordained. There is no inevitable progress leading to intelligent life – leading inevitably to us. The history of life is contingent upon unforeseeable chance events. This should not surprise us, for we know that human history is highly contingent; the life of each of us, and indeed our very birth, is the infinitesimally improbable result of myriads of unlikely chance encounters. Steven J. Gould has stressed this aspect in the history of life by pointing to the bizarre fauna of the 530-million-year-old Burgess Shale. Studying the beautifully preserved soft parts of the Burgess fossils, Harry Whittington, Simon Conway Morris, and Derek Briggs of Cambridge University have discovered a weird bestiary of animals who might well have been the ancestors of life on the planet today, but instead died out. Somehow the pathways of chance led to us rather than to the never-realised descendents of those strange ancient animals. For an explanation of these pathways, see Christopher Zeeman's chapter on evolution and catastrophe theory in this volume.

Impact catastrophes seem to have played another unexpected role in evolution. The fossil record indicates that in normal, quiet times each species becomes better and better adapted for exploiting its own particular ecological niche, and it becomes harder and harder for any other species to evolve in such a way as to take away that niche. The fossil record seems to indicate that the rate of evolution slows during normal times. However, the wholesale removal of species from their niches by a giant impact provides a great opportunity for the survivors, like the discovery of a whole new world to populate. The fossil record shows us that the rate of evolution accelerated immediately after the mass extinction at the end of the Cretaceous, as the surviving species evolved in ways which let them spread into and occupy the vacant niches. We have heard graduate students compare this situation to the excellent job prospects they would face if half of the tenured professors were suddenly fired.

Among the happy survivors of the Cretaceous–Tertiary extinction were our ancestors, the early Tertiary mammals. When dinosaurs were the dominant large animals on Earth, the mammals seem always to have been small and insignificant. However, with the removal of the dinosaurs from the terrestrial scene, the mammals began an explosive phase of evolution, leading eventually to intelligent human beings. We may find ourselves pausing from time to time and reflecting that this was a crucial turning point in the history of life, and that we owe our very existence as thinking human beings to the impact which destroyed the dinosaurs.

## FURTHER READING

Alvarez, L. W., Alvarez, W., Asaro, F. and Michel, H., 'Extraterrestrial cause for the Creta-ceous–Tertiary extinction', *Science*, 208, no. 4448, 1095–1108, 6 June 1980.
Alvarez, W., Hansen, T., Hut, P., Kauffman, E. G. and Shoemaker, E. M. 'Uniformitar-

ianism and the response of earth scientists to the theory of impact crises', *Catastrophes and evolution: astronomical foundations*, Cambridge, Cambridge University Press, 13–24, 1989.

Glen, W. 'What killed the dinosaurs?', *American Scientist*, 78, 354–370, 1990.

Montanari, A., Hay, R. L., Alvarez, W., Asaro, F., Michel, H. V., Alvarez, L. W. and Smit, J., 'Spheroids at the Cretaceous–Tertiary boundary are altered impact droplets of basaltic composition', *Geology*, 11, 668–671, 1983.

Muller, R. A. *Nemesis*, New York: Weidenfeld and Nicolson, 1988.

Raup, D. M. *The Nemesis Affair*, New York: W. W. Norton, 1986.

Smit, J. 'Meteorite impact, extinctions and the Cretaceous–Tertiary boundary', *Geologie en Mijnbouw*, 69, 187–204, 1990.

# Darwin and catastrophism

*MARTIN RUDWICK*

'Nice soft wife on a sofa'; 'better than a dog anyhow'. So calculated the 29-year-old Charles Darwin (1809–82), in his famous double-entry memorandum to himself: to 'Marry' or 'Not [to] Marry'; 'This is the Question'. 'Marry – Marry – Marry. Q.E.D.', Darwin concluded, and soon did so. If we could take a time-machine, and propel ourselves like Dr Who a century and a half into the past, arranging to emerge in London in 1840, we could find Charles and the former Miss Emma Wedgwood ensconced in their house in Upper Gower Street, with their first child (and doubtless a sofa too, but no dog).

Darwin was still busy writing up the scientific results of his trip round the world as unofficial naturalist on H.M.S. *Beagle*. To Robert Fitzroy (1805–65), who had been the ship's captain and Darwin's gentlemanly companion through those five long years, he wrote: 'What I learnt in Natural History [during the voyage] I would not exchange for twice ten thousand [pounds] a year' – a millionaire's fortune in modern terms.

In Darwin's day 'Natural History' covered far more than the collecting of beetles and orchids. It also included geology. 'There is nothing like geology', he had written to his sister Catherine from the Falkland Islands, early in the voyage; 'the pleasure of the first day's partridge shooting or first day's hunting cannot be compared to finding a fine group of fossil bones, which tell their story of former times with

GREAT DECISIONS IN THE
HISTORY OF SCIENCE

*Charles Darwin: "This is the Question"*

*"... nice soft wife on a sofa ..."*

*"... better than a dog anyhow ..."*

*Figure 1* 'Nice soft wife on a sofa; better than a dog anyhow':
a modern interpretation by Pauline Dear of Charles Darwin's
private memorandum on the advantages of married life.

almost a living tongue'. While he was still in South America, the 'geological notices' in the scientific letters he sent back to England began to make him well known in scientific circles. His patron John Henslow, Fellow of St John's College, Cambridge, and Professor of Botany and formerly Professor of Mineralogy in the University, arranged for the Cambridge Philosophical Society to print extracts in a pamphlet, which became Darwin's very first scientific publication. When the voyage was over, and he settled in London to work on his collections, Darwin quickly made the Geological Society of London the focus of his scientific life. He presented all his early papers there, and served as one of its secretaries. So it was as a geologist that Darwin first became known among 'men of science' (as they were called) in London, Cambridge and elsewhere. And this was his own perception too: just before their wedding, for example, he told Emma that everything in their house would be hers too, 'from me, the geologist, to the black sparrows in the garden'.

Among the 'vivid and delightful pictures' from the *Beagle* voyage that were in Darwin's mind when he was writing to Fitzroy a century and a half ago, we can therefore be sure there were many that bore on the geological research from South America that he was now writing up. But he was also hard at work on another topic that likewise stemmed from what he had seen on the voyage. This too he had discussed at the Geological Society soon after his return from the voyage. He was taking his short paper on the geological significance of the coral reefs he had studied in the Pacific, and he was turning it into a book. Only the day before writing to Fitzroy, Darwin sent a letter to his geological mentor Charles Lyell (1797–1875) about this work. (It only had to go to the other side of Bloomsbury: today its contents would have vanished down the telephone.) Darwin told Lyell his latest ideas about coral reefs and atolls, and he called him 'the one man in Europe, whose opinion of the general truth of a longish argument I should always be most anxious to hear'.

Darwin's concern about Lyell's opinion was understandable. Darwin claimed that the various forms of coral reef could act as subtle indicators of vast but very slow vertical movements in the Earth's

*Figure 2* A draft cartoon (1831) by Henry De la Beche, mocking the 'uniformitarian' theories in Lyell's *Principles of Geology*. The original caption read: 'The balance of power – or how to keep the sea at its proper level: "Here we go up, up, up; here we go down, down, down" '.

crust. This emphasis on the slow and gradual nature of such crustal movements was the essence of Lyell's approach to geology. A cartoon by Henry De la Beche (1796–1855), one of Lyell's most acute critics, was meant to ridicule the idea, but at the same time expressed it rather well (see Figure 2). Father Time holds the balance, with Europe and Africa in one pan and America in the other, while timing the movements with a clock marked in 'millions of centuries'. Darwin knew that his theory about coral reefs and atolls provided Lyell with a new (or more precisely, improved) line of evidence in favour of his 'uniformitarian' interpretation of geology as a whole. At the same time, therefore, it also became new evidence to use against Lyell's 'catastrophist' critics.

Those unwieldy terms had been coined a few years earlier by William Whewell (1794–1866), Fellow and Tutor of Trinity College, Cambridge, Professor of Moral Philosophy and omnicompetent knowall. (Some of Whewell's other neologisms have fared better: for example, 'Eocene, Miocene and Pliocene', 'anode and cathode', and above all, 'scientist'.) In a major review of Lyell's famous book the *Principles of Geology* (1830–3), Whewell contrasted Lyell's approach to that of most of his contemporaries. 'These two opinions', he wrote, 'will probably for some time divide the geological world into two sects, which may perhaps be designated as the *Uniformitarians* and the *Catastrophists*.' That sentence launched those polysyllabic labels into the world, and they have served ever since to define a fundamental problem in the earth sciences.

So, after starting out with young Darwin and his 'nice soft wife on a sofa', we reach the real subject of this chapter: Darwin and catastrophism. But the introduction has served to set the scene for understanding a debate of major importance in the history of science. The debate was started by these gentlemanly 'men of science' around the time Queen Victoria came to the throne; but modern earth scientists are now engaged in a replay of much the same arguments.

## UNIFORMITARIANISM

If we read on in Whewell's review of Lyell's book, we get at once a sense of how catastrophism, and not Lyell's and Darwin's alternative, was the way that most good scientists in the 1830s found it most plausible to interpret the past history of our planet. Catastrophism, Whewell wrote, 'has undoubtedly been of late the prevalent doctrine, and we conceive that Mr Lyell will find it a harder task than he appears to contemplate, to overturn this established belief. Indeed, we think it ought to be so.'

Whewell was no lightweight critic for Lyell to have to contend with, nor was he mistaken in claiming that Lyell was in a small minority among competent geologists in the 1830s. While he was President of the Geological Society – it was he who persuaded a reluctant Darwin to serve as Secretary under him – Whewell brought out a massively learned *History of the Inductive Sciences* (1837) in three volumes. Three years later (just as we discovered Darwin and his wife in

Upper Gower Street), he followed this up with an equally profound *Philosophy of the Inductive Sciences*, with the significant subtitle, 'founded upon their history'.

In these volumes on the history and philosophy of science, Whewell attempted to redraw the map of knowledge, classifying the various sciences according to their philosophical foundations. As an admirer of the great German universities, Whewell took the 'sciences' to mean all the *Wissenschaften*: he had no time for what was then an upstart anglophone novelty, by which only the natural sciences were allowed the title 'science'. So we find him putting geology in the same category as philology and archaeology. What bound those natural and human sciences together was their common concern with re-constructing *histories*. But what in his opinion made them all 'philo-sophical' (or in our terms, scientific) was their concern to understand the *causes* of past events. Whewell never missed an opportunity to parade his classical learning; so he combined the Greek roots for 'ancient' and 'cause', and coined the term 'Palaetiological Sciences' as a general umbrella for *all* the sciences that not only try to reconstruct the unobservable past, but also try to explain that past in causal terms.

The word 'palaetiology' never caught on, but it is still important because it highlights what Whewell rightly saw as the fundamental, intrinsic and therefore perennial problem for *any* such science. The problem is this: how can we know about the unobservable past? More specifically, how are we to interpret whatever traces of the past have survived into the present? And how are we to explain the causes of past events, if not in some way in terms of the repertoire of causes that we can observe around us now? And although Whewell put the study of the development of languages and technologies into the same category, it was geology that he thought was leading the way.

What then was the essence of Lyell's 'uniformitarian' geology? How did it contrast with the better established approach that Whewell called 'catastrophism'? Lyell's three-volume *magnum opus* bore a fighting title (Darwin took the first volume with him on·the *Beagle*, and had the other two sent out to him). *Principles of Geology* was no modest phrase for the 33-year-old Lyell to choose. In 1830, any scientific book with the word 'principles' in its title evoked memories of the great Isaac Newton's *Principia*, which was a hard act to follow. Lyell's work was no mere elementary textbook.

Lyell's subtitle, too, was made up of fighting words. The book was to be 'an attempt to explain the former changes of the earth's surface, by reference to causes now in operation'. The word 'attempt' suggests a little modesty, but in fact Lyell made no secret of his firm conviction that the attempt could be, and in due course would be, 100% success-ful. All the phenomena of geology, all the traces of the past history of the earth that geologists studied: all of them, Lyell believed, would yield to proper interpretation. All of them, he was confident, could be explained in terms of processes the same in kind *and* degree as those that could be observed in operation in the modern world, or at least those that were reliably recorded in human history.

Opposite the title page, Lyell placed a frontispiece that summarised this objective in visual terms – it was the equivalent of our eye-catch-ing dust-jackets (Figure 3). Lyell's readers expected a book on geology to have a frontispiece showing some spectacular piece of scenery, or a geological section through the earth's crust. Instead, they found here an engraving of a relic of classical antiquity; Lyell had borrowed it from an Italian antiquarian's study of the region around Naples. But the so-called Temple of Serapis, which Lyell had seen for himself two years earlier, suited him perfectly. About one third the way up, the ruined columns had been bored through by marine organisms that normally do such damage just beneath the tideline. So the ruin was a vivid witness to significant movements of relative sea-level in this area, even within historic times.

More than that, the ruin showed there had been such movements in *both* directions. (You can still see it today, much the same except for the TV aerials on the skyline.) The Temple had presumably been built on what was dry land in Roman times; then at some later time it must have been submerged to half-way up the columns, and evidently for a longish period at that; but then at some still later time it had re-emerged, though not quite back to its original level (so the

*Figure 3* The so-called Temple of Serapis at Pozzuoli near Naples, as depicted in the frontispiece of Lyell's *Principles of Geology* (1830).

columns now stand in shallow sea-water). Multiplied over enough time, through periods unimaginably older than classical antiquity, it became conceivable that such small-scale crustal movements would be adequate even to sink whole continents below the ocean, or conversely to raise the Alps or Andes above the snowline. So Lyell would argue.

As evidence for small-scale non-directional change within human history, the Temple of Serapis became a perfect microcosm for Lyell's concept of Earth history as a whole. He kept it as frontispiece through more than forty years and twelve editions of the *Principles of Geology*. It even came to be embossed in gold on the cover of the book, becoming a kind of secular icon of uniformitarianism. In effect, Lyell claimed that the past history of the Earth had never been fundamentally different from what we can observe about it in the present.

Of course, its continents and oceans, its mountains and deserts, most of all its animals and plants, had not always been the same. Indeed, geology was a spectacularly exciting science at this time, precisely because it showed the huge scale of past changes. Popular books on geology were telling the British reading public that below their familiar countryside were traces of times when that part of the globe had been successively a tropical swamp, a hot desert, and a vast warm ocean. In the cliffs of Lyme Regis, for example, were traces of an 'older Dorset', as De la Beche depicted it in one of the earliest reconstructions of its kind (Figure 4). On the site of the present Dorset there had been a sea full of strange and decidedly unEnglish creatures. The history of the Earth had clearly been nothing if not eventful.

Lyell agreed; but he claimed that all these huge changes had been achieved with no more sudden violence than could be seen in the winter storms on the English coast or the winter torrents in the Highlands; or in the kind of ecological changes that had gradually made the wolf and the bear extinct in Britain within historic times. At most, Lyell would invoke the humanly catastrophic but geologically quite minor effects seen from time to time in less favoured parts of the world: for example, the historic eruptions of Vesuvius, a recent major earthquake in Chile, a terrible flood in Bengal, and so on.

This then was the essence of uniformitarianism. On the one hand, Lyell paraded it as a method of analysis for inferring the past from the present, in what Whewell was soon to term a 'palaetiological' science. Indeed, Lyell claimed it was the *only* proper method for geology, the only method that embodied the fundamental principle of the basic

*Figure 4* De la Beche's humorous – but scientifically reasoned – reconstruction of a 'more ancient Dorset', inhabited by the extinct animals found as fossils in the Jurassic strata of Lyme Regis (1830).

'uniformity of nature', which necessarily underlay *any* genuine science of nature. Only 'causes now in operation', as Lyell put it, were to be admissible as agents in explaining the past. But on the other hand, uniformitarianism became in Lyell's work something much more than a method. By a kind of creative confusion, Lyell mixed his method with what can best be called a vision of Earth history: a vision of an Earth that had always been roughly the same kind of place as it now is in the brief moments of human history. 'Time's stately cycle', Steven Jay Gould dubbed it recently: stately, because Lyell conceived all geological change as gradual; a cycle, because he admitted no overall directional change in the history of the Earth.

That confident vision of 'time's stately cycle' was just what Darwin shared with Lyell. Anyone who saw the shapes of coral reefs through Darwin's eyes realised that. It is no wonder, then, that Darwin quickly became Lyell's closest associate. But in fact, Lyell needed Darwin as a junior ally almost as much as Darwin needed Lyell as a senior patron. The 'uniformitarians' were a small and embattled minority. Whewell had called them a 'sect'; but if the ecclesiastical metaphor were apt, then the 'uniformitarians' were a very small and particular sect indeed.

## CATASTROPHISM

Lyell, too, found theological imagery appropriate to express his situation. In 1832, as he struggled to write the third and culminating volume of the *Principles of Geology*, he told his fiancée, 'I am grappling not with the ordinary arm of flesh, but with principalities and powers, with Sedgwick, Whewell and others, for my rules of philosophizing, as contra-distinguished from them, and I must put on all my armour'. Like St Paul, he saw himself girded against powerful adversaries.

To what, then, was Lyell's vision opposed? What was the essence of catastrophism, and who maintained it against Lyell's indisputably well-argued case? To answer such questions, we have to penetrate the smokescreen of Lyell's persuasive rhetoric. It was not for nothing that he had been trained and called to the Bar at Lincoln's Inn, and even practised as a barrister on the Western Circuit, before turning full-time to geology. In the *Principles of Geology*, Lyell used all the tricks of the barrister's trade, to make easy mincemeat of his scientific opponents. He described the theories of the catastrophists in a way that made them sound deeply unscientific. 'We hear of sudden and violent revolutions of the globe', he thundered,

> of the instantaneous elevation of mountain chains, of parox- ysms of volcanic energy ... We are also told of general catastrophes and a succession of deluges, of the alternation of periods of repose and disorder, of the refrigeration of the globe, of the sudden annihilation of whole races of animals and plants, and other hypotheses, in which we see the ancient spirit of speculation revived, and a desire manifested to cut, rather than patiently to untie, the Gordian knot.

This is the stuff of good courtroom drama, but hardly the place to find out what Lyell's critics really thought. The point needs emphasis, because modern historians and geologists tended until recently to take Lyell's eloquent prose as a fair characterisation of the catastroph- ism of his opponents. But listening to speeches by the prosecution is rarely a good way to find out the case for the defence.

When Lyell told his fiancée about the demonic 'principalities and powers' that he was contending against, two outstanding Cambridge scientists did duty for the catastrophists. Whewell we have already met; but equally well known in scientific circles was Adam Sedgwick (1785–1873), also a Fellow of Trinity College, Cambridge, Wood- wardian Professor of Geology, and a perennial hypochondriac. It was Sedgwick, perhaps the best field geologist in England, who had given the young Darwin a crash course in fieldwork just before he set sail on the *Beagle*, and so started him on his geological career. But Sedgwick had also been President of the Geological Society when the first volume of Lyell's *Principles of Geology* was published. Sedgwick did not have Lyell's advantage of being trained in courtroom oratory; but he was experienced in another kind of oratory, that of the pulpit. So he was well able to hold his own against Lyell's arguments. In fact, the foundations of catastrophism were never so clearly set out as when Lyell's attack forced catastrophists to make them explicit.

Right at the outset, Sedgwick protested at Lyell's appropriation of the 'principle of uniformity'. He insisted that catastrophists had as much right to it as unformitarians. He told the assembled geologists (who of course included Lyell) that he too believed firmly in the uniformity of all the fundamental processes of physical nature. In that sense, of course Nature was uniform. But Lyell wanted much more than that: he claimed that there must also be 'uniformity' in far more complex processes such as volcanic action and earthquakes, and particularly, uniformity in the intensity of those processes throughout Earth history. 'This theory', proclaimed Sedgwick from his presidential pulpit,

> confounds the immutable and primary laws of matter with the mutable results arising out of their irregular combination. It assumes, that in the laboratory of nature, no elements have ever been brought together which we ourselves have not seen combined; that no forces have been developed by their combination, of which we have not witnessed the effects. And what is this but to limit the riches of the kingdoms of nature by the poverty of our own knowledge; and to surrender ourselves to a mischievous, but not uncommon philosophical scepticism, which makes us deny the reality of what we have not seen, and doubt the truth of what we do not perfectly comprehend.

Lyell's rigorous definition of uniformity, Sedgwick concluded, was 'an unwarrantable hypothesis with no *a priori* probability, and only to be maintained by an appeal to geological phenomena'. There was the rub: Sedgwick and his friends believed that the plain evidence to be seen in the field demanded explanation in terms of occasional episodes of great violence. Lyell, they argued, had to explain that evidence *away*, by assuming, for example, that there were vast unrecorded tracts of time that created an illusion of sudden violence. The catastrophists, not the uniformitarians, claimed to be the plain no-nonsense empiricists in this debate. Once again, a cartoon by De la Beche makes the point perfectly (Figure 5). Lyell the barrister, his

*Figure 5* De la Beche's draft cartoon (from the same series as
Figure 2) mocking the theory-laden character of Lyell's geology.

critics felt, was offering them the coloured spectacles of interpretation, which distorted the plain sense of their observations; the distor-
tion was due to the *theory* embodied in the viewpoint Lyell stood on,
and in his book that he was concealing behind his back.

Conveniently for Sedgwick, a fine example of catastrophist theorising, every bit as ambitious and as well grounded as Lyell's, had just
been published, not in England but in France; and Sedgwick turned
to it, in pointed contrast, immediately after disposing of Lyell. 'In
reading the admirable researches of M. [Elie] de Beaumont', he said,
'I appeared to myself, page after page, to be acquiring a new geological sense, and a new faculty of induction'.

Léonce Elie de Beaumont (1798–1874), one of the best geologists in
Europe, had synthesised a vast mass of fieldwork evidence from all
over the world, and come up with a grand theory about mountain
building (1829–30). Geologists had suspected for several years that
not all mountain chains were of the same age. But Elie de Beaumont
showed how they could be dated quite precisely: not of course in
years or millions of years, but on the geologists' relative time scale.
One had to find out by fieldwork what were the youngest formations

of strata that had been deformed by the mountain-building move-ments, but also what were the oldest formations that were undis-turbed. Then it followed that that particular mountain chain must have been elevated in the interval between the deposition of those two formations. Elie de Beaumont presented this argument diagram-matically in the form of a long fold-out geological section (Figure 6). From left to right he depicted a procession of successive mountain ranges; each one had deformed or twisted upwards a formation that had *not* been deformed by the previous one to the left.

This then became the basis for a powerful theory. The long-term history of the Earth, Elie de Beaumont argued, had been punctuated by occasional episodes of mountain-building. Each time the Earth's crust had buckled and deformed all the then-existing strata, at least locally. After each such episode, there had been another very long period of quiet, during which mountains and hills were slowly worn down just as we see happening today; and the sediment accumulated slowly to form new strata, often lying 'unconformably', as geologists say, on the eroded edges of older formations. Then, in the next vio-lent episode, those strata too, along with all the older ones, would be deformed in their turn.

In the days before geologists could use radioactivity to get quanti-tative estimates of ages, they had no option but to talk in qualitative terms. But Elie de Beaumont evidently thought these episodes had been very sudden and very violent in relation to the vast tracts of time between. And the end of his diagram that depicted the most recent epochs makes it clear that he thought these mountain-building epi-sodes had continued to punctuate Earth history right up to the present; or at least, the geologically recent past. He suggested ten-tatively that the most recent episode might have been the elevation of the Andes.

Elie de Beaumont's theory is a good example of catastrophist theorising in Darwin's day. Certainly it invoked rare events of a kind that had never been recorded in human history. But if these events really were rare, why (said the catastrophists) should we expect that human beings would ever have witnessed one? Certainly the theory

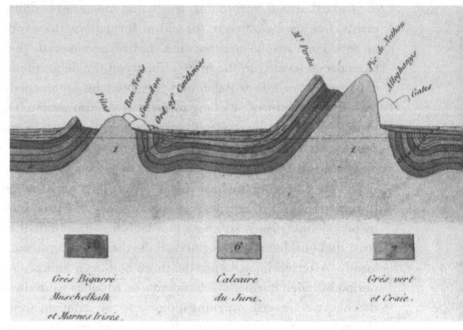

*Figure 6* Part of Elie de Beaumont's highly schematic geological section (1829–30), drawn to illustrate his catastrophist theory of the occasional sudden elevation of mountain ranges (time is represented as flowing from left to right).

invoked a process of enormous magnitude. But it did so (said the catastrophists) for the good Newtonian reason that postulated causes must always be commensurate with the effects to be explained: huge mountain ranges demand huge forces to elevate them. And contrary to what Lyell insinuated, the catastrophists were not smuggling in 'extraordinary' causes, let alone supernatural ones, in any unscientific way.

In fact, Elie de Beaumont argued from the best geophysics of his time to suggest a possible cause of an impeccably natural and physical kind. Most physicists and most geologists in the 1830s agreed that the Earth was probably cooling slowly from an originally hot fluid state: the idea had all the respectability that came from its endorsement by the great French mathematician and physicist Joseph Fourier. And they believed a large part of the interior was still fluid

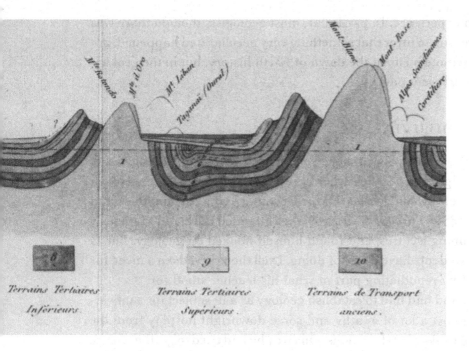

Terrains Tertiaires Inférieurs.   Terrains Tertiaires Supérieurs.   Terrains de Transport anciens.

(hence the outpourings of molten lava in volcanos). Elie de Beaumont adopted this geophysical theory, and suggested that as the fluid core slowly cooled and shrank over the immeasurable aeons of Earth history, the solid crust would build up increasing strains until – suddenly – it cracked and buckled upwards. Then there would be another long stable period, until the strains had built up again, only to be relieved by another sudden buckling of the crust, in different places and in different directions.

Elie de Beaumont's theory was nothing if not impeccably naturalistic. And equally, it was not constricted by any shortage of geological time, because it postulated vast tracts of tranquil time *between* the occasional violent episodes. This then was the kind of theory that catastrophist geologists found plausible; certainly more so than Lyell's vision of 'time's stately cycle'. Earth history had unquestionably been immensely long, in fact quite literally, unimaginably long. But they reckoned there was good evidence that it had been punctuated by occasional episodes of kinds that had no close parallel in human experience. In particular, most geologists thought there was good reason to infer that something very peculiar had happened, not in the remote mists of the dawn of Earth history, but in the geologically very recent past.

## DILUVIALISM

Just as Darwin had been started on a geological career by Sedgwick at Cambridge, so a decade earlier Lyell had been started on *his* geological career by Sedgwick's Oxford counterpart, William Buckland (1784–1856). Three years in a row Lyell attended Buckland's famously entertaining lectures, and found himself hooked on geology. But as bright students have a way of doing, Lyell then carved out a niche for himself by repudiating part of what his teacher stood for.

Buckland had had to establish geology as a fit subject for a university, against a lot of apathy and some downright hostility from the massed ranks of Oxford dons. One way he had tried to get the science accepted was to emphasise that its results were quite compatible with traditional 'sound learning'. His inaugural lecture in 1819 promised his donnish audience that 'the Connexion of Geology with Religion' would be explained, and geology thereby vindicated. Specifically, Buckland argued that the results of geological research did not undermine, but actually reinforced, a rather traditional way of interpreting the Bible. Buckland was no fundamentalist (to use the modern term). But when he explained how geologists believed they had evidence of a very unusual event in the geologically recent past, he identified this event as the one that was briefly recorded in Genesis as Noah's Flood. In earlier centuries the Flood or Deluge had often been a favourite way to explain almost anything one wanted to explain about the surface of the Earth, and its rocks and fossils. But Buckland was one of those who brought it back to a new respectability in the early nineteenth century, when it was often made into another '-ism', 'diluvialism'. The diluvial theory became another striking example of catastrophism.

Certainly, Buckland was not alone in thinking that something very peculiar had happened at some remotely prehistoric time, but still very recently in geological terms. Buckland's own research, for example, showed how the English Midlands were strewn with gravels containing chunks of rocks quite different from those known anywhere nearby, and with a very peculiar distribution. They were strewn over the countryside in a way that seemed totally unrelated either to the underlying rocks or to the present drainage system of the area. What could have caused such a distribution? To Buckland, and to most other geologists, it seemed self-evident that the cause could *not* be the present rivers. Instead, they thought these peculiar gravels must have been swept into their present positions by a sudden flood of water, coming in this case from the Midland plain and sweeping south over the low watersheds through the north Cotswolds into the Thames valley.

Sensational *new* evidence for such a 'geological deluge', as it was called, came literally into Buckland's hands not long after his inaugural lecture. As the leading specialist on fossils in England, Buckland was sent some fossil bones from a recently discovered cave in Yorkshire. He announced his sensational results to the Royal Society as soon as he could, and that august body was so impressed that Buckland was awarded its Copley Medal, the first time it had been given to a geologist. From a detailed study of the fossil bones, Buckland inferred that the cave had been a den of hyaenas, who had scavenged the carcases of other animals, dragged them into the cave, and then chewed them up at leisure. This reconstruction was based impeccably on the kind of comparison between present and past that his student Lyell was later to champion as the hallmark of his own work. Buckland took the trouble to study living hyaenas in a zoo, and he showed that they broke and chewed their bones in just the same way as the fossil bones had been broken and chewed.

Buckland's reconstruction was so vividly convincing that his contemporaries felt it was as if he had been there at the time. One of them drew a cartoon to make the point: Buckland was shown crawling into the Yorkshire cave, candle in hand, and there finding the extinct hyaenas still very much alive (Figure 7). But how had the hyaenas

*Figure 7* A cartoon (1822) by Buckland's friend William Cony-beare, showing Buckland entering the Yorkshire cave he had reconstructed as a den of extinct hyaenas dating from *before* the catastrophic 'geological Deluge'.

become extinct, at least in England, along with the animals they had scavenged – tigers, elephants, rhinos, hippos and suchlike unEnglish game? Buckland linked their apparently sudden disappearance with the other 'diluvial' evidence, and in 1823 he published an enlarged version of his Royal Society paper in book form. A huge and violent deluge, he argued, must have swept suddenly and briefly over the whole country, roughly from north to south, wiping out the exotic fauna and flora, hyaenas and all, and sweeping rocky debris even over watersheds in the drainage system. The cave bones became, in this sense, 'relics of the deluge' – the title of his book.

Such an alleged event was nothing if not catastrophic. And as Buckland reviewed the evidence from caves all over Europe, it became clear to his readers that this might have been a catastrophe of global proportions. But it was not left without a possible natural cause. To catastrophist geologists, the event they inferred seemed not unlike the sudden and catastrophic tidal waves that sometimes overwhelmed coastal areas. The famous Lisbon earthquake of 1755 had been accompanied by just such a destructive tidal wave. It seemed conceivable that a far larger earthquake in the remote prehistoric past, or perhaps the abrupt elevation of a new mountain range à la Elie de Beaumont, might have caused a correspondingly larger tidal wave: a mega-tsunami, as it were, to use the modern geologists' term.

If this had been the Deluge, it had certainly *not* been much like what a literal reading of Genesis would suggest. Buckland may have had his own reasons for emphasising the Genesis link, in the context of Oxford university politics. But he was certainly being no simpleminded literalist about it. If the 'geological Deluge' was indeed the same event as Noah's Flood, then the account in Genesis was a pretty poor and garbled record, and it had to be interpreted in a very unliteral fashion. Still, even Buckland's loose link with biblical exegesis was enough to make the severely rationalistic Lyell see red. He set himself to prove that no abnormal geological event had disturbed the orderly peace of Europe in geologically recent times. There was, he argued in effect, no natural analogue to the French Revolution, lurking back in the recent past.

## GLACIALISM

However, in the years that followed Buckland's cave research, the evidence that had made him and others infer some kind of extraordinary event refused to go away. If anything, it became more compelling rather than less. How, for example, could one explain blocks of rock the size of houses, found tens or even hundreds of miles from the nearest source of that kind of rock, and perched on something quite different? What 'cause now in operation' could have transported such huge 'erratic blocks'?

The answer came, just around the time we first visited the Darwins in Upper Gower Street. In the 1830s, several Swiss geologists made a careful study of present-day glaciers in the Alps. They showed that

*Figure 8* Jean de Charpentier's illustration (1841) of a large erratic block, interpreted by him as evidence for the former vast extension of the Alpine glaciers.

glaciers always contain embedded chunks of rock of all sizes, which have fallen from the slopes above the glacier or been plucked from its sides or the underlying floor. As the glacier grinds its way slowly down the valley, these rocks act like a gigantic piece of coarse sandpaper, and scratch the solid rock over which the ice is moving. At the snout of the glacier, where the ice is melting, the embedded blocks of rock are just dropped. And where a glacier has retreated, the scratched bedrock is exposed to view (Figure 9). The Swiss geologists then applied their knowledge of modern glaciers to the understanding of the past, in good Lyellian fashion. But they came to a most unLyellian conclusion.

They found they could trace the erratic blocks and the polished rock surfaces further and further from the existing glaciers. One of them, Jean de Charpentier (1786–1855), traced the distribution of erratics of a distinctive rock, which could be found *in situ* in the mountains above the Rhône valley in the Swiss Alps. The present-day glaciers lie high up in some of the side valleys. But Charpentier found the erratics all over the lower valley, down past the Lake of Geneva, out across the broad Swiss Plain, and even half-way up the hills of the Jura to the

*Figure 9* Louis Agassiz's illustration (1840) of an example of the scratched rock surfaces found far beyond the present Alpine glaciers, and interpreted by him as evidence of an 'Ice Age' of possibly global extent.

north. Charpentier then made the bold inference that the modern glaciers above the Rhône valley, and presumably all the others in the Alps too, had formerly extended far beyond their present limits, far beyond even the Alps themselves, and out over the surrounding lower ground.

This was a theory tailor-made to appeal to the catastrophist mainstream of the geological community. It confirmed that in the geologically recent past there had indeed been an event of dramatically different character from the present. But not after all a deluge: instead, an Ice Age. Another '-ism' entered the geological world: 'glacialism'. The glacial theory became the best and strongest case for catastrophism. Buckland was quickly convinced, and became the most ardent 'glacialist' in England. In 1840, he and another Swiss glacialist, Louis Agassiz (1807–73), toured the Scottish highlands together, and found the evidence of glaciers everywhere. The Ice Age had evidently been more drastic than even the glacialists had expected: not just the area around the Alps, but most of Britain and

Scandinavia too, had apparently been covered in ice. Buckland had planned to write a sequel to his book on the geological evidence for the Deluge, working out a possible causal explanation for that event. But now he abandoned that plan, because he reckoned the true cause had been found: an Ice Age.

Lyell had a much harder time with glacialism. But as with other kinds of evidence, he managed, if not quite to explain it away, at least to modify it to make it less threatening to his uniformitarian scheme of things. He conceded that there had indeed been a colder period in the recent past, but he argued that erratic blocks must have been shipped to their present positions on icebergs. To be fair to Lyell, this was just the time when the polar regions were being thoroughly explored for the first time, and his suggestion was not at all far-fetched: Darwin supported him by publishing a sailor's sketch of an iceberg carrying a 12-foot boulder in the Antarctic.

The long-term significance of glacialism was that it became the best and clearest cause of a catastrophist kind of theory. The idea of an Ice Age was certainly based on inference from present-day glaciers and glacial action. But it postulated that at one particular time in the past, and with great suddenness (at least in geological terms), those effects had acted with far greater intensity than at present. The ice-sheets had not been safely confined to Greenland and Antarctica, but had crept even over Oxford and Cambridge, down to the outskirts of London. The Ice Age was indeed a 'catastrophe' in the geological sense, and it showed the limitations of Lyell's rigorously 'uniformitarian' geology. Nonetheless, despite the striking character of the Ice Age, a watered-down version of Lyell's vision of Earth history came to dominate geologists' thinking in the later nineteenth century, and well into the twentieth; catastrophism became a dirty word.

## CONCLUSION

What, in conclusion, are we to make of catastrophism in Darwin's day? We have seen how the sort of theorising that Whewell called 'catastrophist' was at that time part of the mainstream of good scientific thinking about the Earth; even, indeed, that it *was* the mainstream. It was not itself a single theory; rather it was a style of theorising. Most of the best earth scientists of the early nineteenth century felt that the evidence available to them obliged them to be in some sense catastrophists. They felt that in trying to explain *some* phenomena they had no option but to invoke causal agencies far greater in intensity than those we can now see at work in the world around us, or even those of which we have reliable human records. But they also felt compelled by the evidence to infer that some of these geological 'causes' had acted with such intensity only on very rare occasions in Earth history. Such events, they concluded, fully deserved to be termed 'catastrophes', and they themselves were content to be known as 'catastrophists'.

Like Lyell, the catastrophists tried as far as possible to reconstruct the unobservable past in the light of the observable present. But unlike Lyell, they were sceptical whether that policy would always be adequate. They refused, as Sedgwick had put it, 'to limit the kingdoms of nature by the poverty of our own knowledge'. Lyell's uniformitarianism has been summarised for generations of students in the aphorism, 'the present is the key to the past'. But the catastrophists said, in effect, that far from opening wide a door on to the past history of the Earth, Lyell's prescription for geology would limit it to the constricted vision of a mere key-hole.

This defence of the catastrophists has unavoidably made Lyell look like the chief baddie in the story. But of course there are no goodies and baddies in the real history of science. Lyell was a very great geologist, and his uniformitarian method has been, and remains, an indispensable discipline in the science. But equally indispensable is the catastrophists' open-mindedness about what the past may really have been like, and their willingness to consider the possibility of even the strangest kinds of past, if the surviving evidence seems to demand it.

Back in the 1960s, a handful of historians of geology began trying to rehabilitate the catastrophists, by showing they had been just as scientific as the uniformitarians. What seems astonishing now is that we were bitterly attacked, and even accused of wanting to smuggle supernaturalism back into the earth sciences. Catastrophism was still a dirty word among scientists, and even among historians of science. But since then, the world of science has changed dramatically. The space programme and the lunar landings altered scientific perceptions of the Earth quite radically. Instead of treating the Earth *de facto* as a self-contained system, geologists came to think of it as being indeed a planet among other planets in space. For example, the claim that some terrestrial features are extra-terrestrial impact craters changed quite rapidly from being the opinion of a few marginal figures to being respectable mainstream science.

By 1980, the suggestion that a major episode of mass extinction might have been due to the rare impact of a wandering asteroid or comet could be put forward in the highly respectable pages of the magazine *Science*. Such claims were highly controversial – and they remain so a decade later – but at least eminent senior scientists no longer need to worry about the dangerous resurgence of catastrophism. Whether Lyell and the young Darwin would have approved, had they been able to read Walter Alvarez and Frank Asaro's contribution to this volume, is of course quite another matter. But I think those two great Cambridge scientists, the Reverend Professors Whewell and Sedgwick, would have been well satisfied in retrospect with their own contribution to the progress of geological science.

## FURTHER READING

Gould, Stephen Jay, *Time's Arrow, Time's Cycle*, Cambridge, Massachusetts: Harvard University Press, 1987.

Greene, Mott T., *Geology in the Nineteenth Century*, Ithaca, New York: Cornell University Press, 1982.

[Herries-] Davies, Gordon L. *The Earth in Decay*, London, Macdonald, 1969.

Rudwick, Martin J. S. *The Meaning of Fossils*, University of Chicago Press, 1984.

# 4

# Evolution and catastrophe theory

*CHRISTOPHER ZEEMAN*

Catastrophe theory is the name given to a method of mathematical modelling introduced by René Thom in the 1960s. It is based on deep theorems in topology, and is particularly applicable to phenomena in which continuous causes produce discontinuous effects. Without a model such effects can be unexpected, and if they happen to be harmful are liable to be called catastrophes – hence the name. On the other hand, with an appropriate model the effects can be understood and anticipated. The easiest way to explain the method is to describe a particular example, and so in honour of Darwin College I have chosen an application to Darwinian evolution.

Charles Darwin assumed that the causes of evolution were gradual: small random variations and natural selection. Yet some of the observed effects appear to be discontinuous. As a result there has been continual debate over the matter ever since he published his famous book *The Origin of Species* in 1859. Indeed, Darwin himself was worried about it and devoted a whole chapter of the book to 'difficulties on theory'. And his strongest supporter T. H. Huxley wrote to him the day before the book was published warning him: 'You have loaded yourself with an unnecessary difficulty in adopting *Natura non facit saltum* (nature does not make jumps) so unreservedly'. Even today the argument continues: for instance, Niles Eldredge and Stephen Jay Gould have introduced the term *punctuated equilibria* to describe the fact that species tend to remain stable for a long time, but if one species evolves into another it often does so relatively rapidly. Other biologists emphasise Darwin's gradualism, and dismiss punctuated equilibria by saying that of course there are likely to be varying rates of evolution but nothing to make a fuss about.

We shall reconcile the two points of view by making a catastrophe model of continuous cause and discontinuous effect as follows. We translate Darwin's gradualist hypotheses into mathematics in the simplest possible way, and then deduce from the mathematics that punctuated equilibria is not a contradiction to, but a *consequence of*, those hypotheses.

Furthermore, we shall explain why *multiple speciation* is likely to occur at the punctuation points. Here, multiple speciation means the appearance of several descendant species. Both phenomena, punctuated equilibrium and multiple speciation, are observed in the fossil record. Some evolutionary theorists suggest that multiple speciation is not 'generic' and should be replaced by a sequence of bifurcations, but the mathematics suggests the opposite, vindicating Darwin's original diagram (in fact the only diagram) in *The Origin of Species* (Figure 1, left).

Catastrophe theory has been applied to evolution by Maurice Dodson and a number of other authors, but this is the first explanation of multiple speciation.

Before getting down to the business of making the model let us review in more detail the observed discontinuities in the evolution of species that the model ought to explain. There are three types:

> discontinuities in time (punctuated equilibria);
> discontinuities in space (abrupt frontiers); and
> discontinuities in form (speciation, canalisation).

## DISCONTINUITIES IN TIME

These can be observed in the fossil record. A typical example is a surface parallel to the rock strata, below which the strata contain fossils

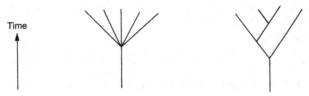

*Figure 1* Left: multiple speciation. Right: bifurcations.

of a species A, and above which the species A has been replaced by a species B (see Figure 2, left). The corresponding point on the geological timescale is called a *punctuation point*. This sudden disappearance of A at the punctuation point and its replacement by B is usually interpreted as B invading A's territory and occupying the same ecological niche that was previously held by A.

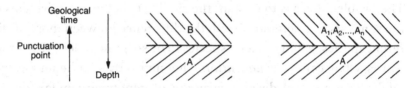

*Figure 2* Left: invasion. Right: multiple speciation.

Another typical example is the disappearance of A and its replacement by one or more descendant species $A_1, A_2 \ldots A_n$ (see Figure 2, right). The interpretation here is of a sudden evolution at the punctuation point, with multiple speciation. The model will explain both processes.

## DISCONTINUITIES IN SPACE

The above examples are of discontinuities in time at a fixed point in space. Similarly these are also discontinuities in space at a fixed point of time, and these can be observed today as abrupt frontiers between different species occupying the same ecological niche (Figure 3).

A familiar example is the tree line on a mountain, above which trees are replaced by grass or other plants. In *The Origin of Species*, Darwin draws attention to several examples – going up mountains, down an ocean shelf, or towards cooler latitudes. In the chapter on

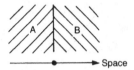

*Figure 3* An abrupt frontier between species.

the difficulties of the theory he argues how intermediate species might be eliminated, but does not explain how such a frontier could form in the first place. Our model will show how a gradual variation of environment can cause the creation of abrupt frontiers.

## DISCONTINUITIES IN FORM

The problem here is to explain the similarity between individuals of the same species compared with the difference between those of different species. What causes a species to be so homogeneous? In other words what causes *canalisation*, the similarity of phenotype (the adults of a species) despite variation of genotype; similar-looking adults can have surprisingly different genes. The immediate answer is natural selection acting on the phenotype. Therefore let us rephrase the question more sharply: if humans and chimpanzees are both descended from a common ancestor about 6 million years ago why is there not a continuous spectrum of individuals running from human form to chimpanzee form? The answer is, on the contrary, that there is indeed such a continuous spectrum in form–time space, going back from the human to the common ancestor, and up again to the chimpanzee, as shown in Figure 4. Here, the space of theoretically possible forms is sketched as two-dimensional whereas it should of course be highly multidimensional.

This answer, however, merely throws the question back from the global to the local: how did the bifurcation occur in the first place? To put it another way, what causes *speciation*, the formation of different species from a common ancestor? If one replies 'natural selection' this raises the awkward paradox that we have now invoked the same cause for the opposite processes of canalisation and bifurcation. The

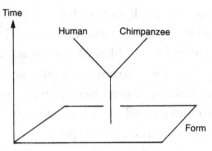

*Figure 4* Form-time space.

model will resolve this paradox, and also explain the alternative process of catastrophe and multiple speciation at punctuation points.

## MATHEMATICAL MODEL

Darwin's theory of evolution can be summarised briefly as:

(1)  Random small variations $\left.\vphantom{\begin{array}{c}a\\b\end{array}}\right\} \Rightarrow$ (3) Evolution of species
(2)  Natural selection

Here (1) and (2) are the local slow gradual cause while (3) is the global long-term effect. We want firstly to translate (1) and (2) into mathematical hypotheses, secondly to derive mathematical results from those hypotheses, and thirdly to interpret the results back into biological conclusions. It is a good idea to keep the mathematical argument as separate as possible from the two modelling processes of translation and interpretation, so that if the biological conclusions turn out to be wrong then the fault can be traced back to one or other of the two modelling processes. So let us begin to translate (1) and (2) into mathematics.

'Small variations' suggests representing the individuals of a species by points in some $n$-dimensional Euclidean space $X$ of theoretically possible forms. What do the coordinates of $X$ represent? They might be particular measurements significant for the specific example of evolution under consideration, such as length of leg or size of beak, or the concentration of some enzyme in some organ, or the timing of some embryological event that affects the eventual shape of the phenotype. We can either choose $n$ small and confine ourselves to a few explicit variables, or else choose $n$ large and allow ourselves an arbitrary number of implicit variables. Although variations in the phenotype must ultimately arise from variations in the genotype, which are of a combinational nature, nevertheless in many cases a continuous model of the phenotype will be adequate, and often easier. Also natural selection acts on the phenotype.

If, further, we want to model the evolution of a symbiotic relationship, or a prey–predator relationship, or an ecology, then a point of $X$ can represent the symbiotic pair, or the prey–predator pair, or a state of the entire ecology.

We translate the word 'random' into mathematics by assuming that the offspring of an individual $x$ lie in a small neighbourhood of $x$. The size and shape of such neighbourhoods will be important for some applications, but we shall not need to consider them at this stage.

Darwin's hypothesis (2) of natural selection means that more offspring are produced than can survive, so that only those that are fitted to a given environment will survive. The simplest way to translate this into mathematics is to postulate an *unfitness function* $u : X \rightarrow R$, where R denotes the real numbers, and $u(x) < u(y)$ means $x$ is fitter than $y$. (We use unfitness rather than fitness in order to exploit the intuition of a ball rolling downhill.) Let us postpone discussion on the existence and measurability of $u$ until we have begun using it, so that the usage can then indicate what needs to be discussed.

For simplicity suppose at first that $X$ is one-dimensional, and consider the graph of $u$ shown in Figure 5. (At the back of our minds we continue to think of $X$ as $n$-dimensional, and later it will be important to have $X$ at least two-dimensional.) Suppose that the individuals of a species have mean $x$, and are clustered in a neighbourhood of $x$. By hypothesis (1) their offspring will be spread over a slightly larger neighbourhood, and by hypothesis (2) only the fitter will survive, and so the mean of the survivors will be slightly downhill from $x$. Hence the species will begin to evolve by rolling downhill towards the minimum A of $u$. When the species reaches A then it will stabilise, because any small variations will be less fit, and so will not survive.

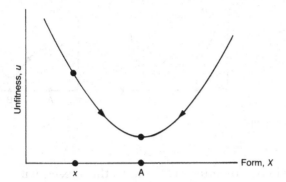

*Figure 5* Unfitness function.

Thus a minimum of $u$ represents *a species in stable equilibrium*. Moreover we have trivially explained canalisation, the similarity of phenotype despite variation of genotype.

It is tempting to think of evolution as a gradient differential equation, but we shall see later that if $\dim X \geqslant 2$ then it obeys a rather different kind of dynamics.

## EXPLANATION OF DISCONTINUITIES IN TIME

We now introduce elementary notions of catastrophe theory. Suppose that the environment is gradually changing so that another ecological niche appears at B, as in Figure 6(*a*). More precisely, assume that $u$ gradually changes so as to create a new minimum at B, which represents a theoretical form of phenotype best fitted to fill that new ecological niche.

Even if the niche gradually becomes more advantageous than A, with a lower minimum as in Figure 6(*b*), the species is prevented from evolving to B because it cannot climb over the intervening hump towards B; any random small variation of individuals that happened to be towards B would immediately be eliminated because they would be less fit than the existing population at A. This situation persists until the gradual change in $u$ causes the minimum at A to coalesce with the maximum, as in Figure 6(*c*). Then any random small vari-

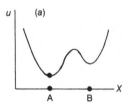 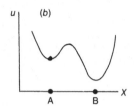 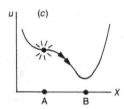

*Figure 6* Catastrophe.

ation towards B will be advantageous, and so the species will evolve until it reaches B, where it will stabilise again. Compared with the periods of stability previously spent at A, and subsequently spent at B, the time taken to go from A to B is likely to be relatively short, perhaps as little as one thousandth as long as the periods of stability. Hence, on the geological timescale it will appear as a catastrophic jump, or more briefly a *catastrophe*. In the fossil record, the catastrophe will occupy a thin layer of perhaps one thousandth of the thickness of the strata below containing fossils of A, and the strata above containing fossils of B. Indeed the layer may be so thin as not to contain any surviving fossils of the intervening forms between A and B, and hence it will appear as a surface of discontinuity at a punctuation point.

Thus we have shown that *Darwin's two local hypotheses imply punctuated equilibria*. Indeed, Darwin himself had insight into this process because in the summary of Chapter X of the first edition of *The Origin of Species* he writes: 'the duration of each formation is, perhaps, short compared with the average duration of specific forms'; and in later editions he expands this to 'although each species must have passed through numerous transitional stages, it is probable that the periods, during which each underwent modification, though many and long as measured by years, have been short in comparison with the periods during which each remained in an unchanged condition'.

We are not claiming that all evolution takes place by catastrophes; the model also allows for gradual evolution. Figure 7 illustrates the two possibilities, beginning and finishing with the same *u*. In each

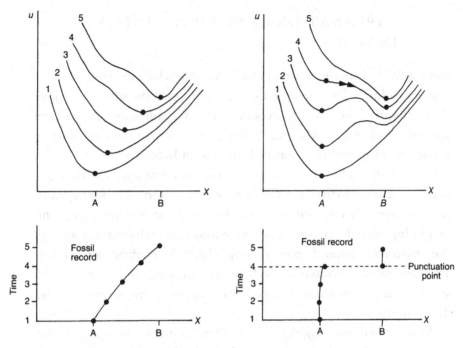

*Figure 7* Left: gradual evolution. Right: punctuated equilibria.

case the gradual change of $u$ is illustrated at successive times $1 \ldots 5$, and the resulting fossil record in form–time space is indicated below.

## DIGRESSION ON UNFITNESS FUNCTIONS

We have been implicitly using unfitness functions ranging over large changes of form and long periods of time. Are we justified in doing this? The answer is yes for the following reasons. Firstly, fitness can be measured locally between individuals living in roughly the same place at the same time in the same environment, for example by counting the numbers of surviving offspring. This determines local unfitness functions, which can then be glued together to give a global unfitness function as required. The glueing together process is nontrivial mathematically, but can be done thanks to a theorem that André Haefliger proved in his doctoral thesis in 1958.

CHRISTOPHER ZEEMAN

## EXPLANATION OF DISCONTINUITIES
## IN SPACE

Suppose that a species S occupies a domain in which the environment varies gradually from one end of the domain to the other. Suppose that one end of the domain favours a smooth gradual evolution of the species into form A, while the other end favours a smooth gradual evolution into form B. What will happen in between?

Figure 8 shows how to translate this question into a geometric problem of extending to the interior of a cube (or a rectangular box) a graph that has already been given on the boundary of the cube. Here the two independent variables are space and time, where *space* means the one-dimensional direction along which the environment varies. The dependent variable is the form $x$, which lies in a multidimensional space of possible forms $X$ (drawn for convenience as one-dimensional).

We want to draw the graph of $x$ as a function of space and time. We are given the boundary of the graph on the faces of the cube as follows. At the beginning of the time period under consideration the graph is constant, $x = S$, represented by the line on the bottom of the cube parallel to the space axis. At the end of the space axis nearest the reader the graph shows a smooth evolution from S to A, represented by the curved line on the front of the cube. At the other end of the space axis the smooth evolution from S to B is represented by another curved line on the back of the cube. At the end of the time period the space is occupied by the two species A and B meeting at an abrupt frontier F, represented by the dotted line of discontinuity on the top of the cube. The problem is how to fill in the surface of the graph inside the cube. Figure 9 illustrates a solution to the problem, and after we have described it we shall explain why it is *the* solution.

The smooth folded surface in Figure 9 is called a *cusp catastrophe*. When the surface is projected (horizontally) onto space–time the images of the fold curves, which are shown dashed, form a cusp at K; hence the name. Examples of the unfitness function $u$ at four typical points of space–time are illustrated on the left. For points outside the

92

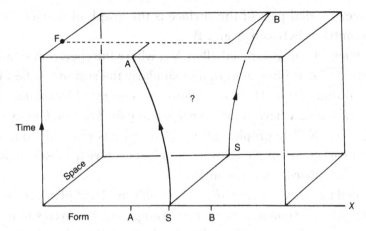

*Figure 8* Graph of form as a function of space–time.

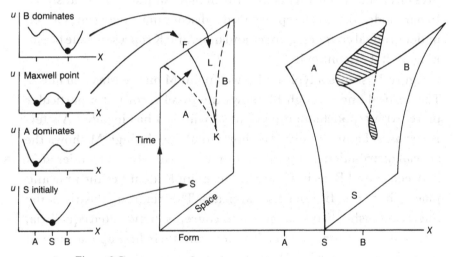

*Figure 9* Cusp catastrophe.

cusp *u* has only one minimum, which at the beginning is S. For points inside the cusp *u* has two minima near A and B, separated by a maximum (or a saddle-point if *X* is multi-dimensional). The two sheets of the surface labelled A and B are the graphs of the minima near the species A and B, and both these sheets emerge smoothly from the lower part of the surface, labelled S, corresponding to the smooth evolution of S into A and B at opposite ends of the space axis. The folded over shaded part of the surface is the graph of maxima (or saddle-points) in-between A and B.

The region of space–time labelled A is where the species A dominates (because A is fitter than B) and similarly the region labelled B is where B dominates. The two regions are separated by the line L of *Maxwell points* where A and B are equally fit. The line L ends at the cusp point K. The graph that we want consists of the dominant species at each point of space–time, and is therefore a subset of the surface, with a discontinuity along L.

L is called the *Maxwell line* after Maxwell's model of phase transition, which has a similar picture. For example, in Maxwell's model the line L represents the discontinuities between liquid and gas of boiling and condensation under gradual changes of temperature and pressure, and K represents the critical point of the phase-transition. In our model, the line L represents the discontinuity between A and B under gradual changes of space and time, and K represents the beginning of that discontinuity.

Figure 10 shows a typical Maxwell point M on the Maxwell line L. The vertical line through M shows the fossil record at a particular place, with a punctuation point at M where B has invaded A's territory as in Figure 2 (left). The horizontal line through M shows the spacial distribution at a particular time, with an abrupt frontier at M between A and B, as in Figure 3. The point K identifies the time and place where this frontier first appears. The diagram illustrates the interrelationship between these four concepts: punctuated equilibria, invasion of territory, abrupt frontiers and their first appearance. It would be interesting if a Maxwell line could be found experimentally by plotting the punctuation points in fossil records at different places.

## DIGRESSION ON THE UNIVERSALITY
## OF THE CUSP CATASTROPHE

Suppose $u$ is any smooth generic function on an $n$-dimensional space $X$ with a two-dimensional parameter space P (such as the space–time plane in Figures 9 and 10). The set C of critical points of $u$ (that is the

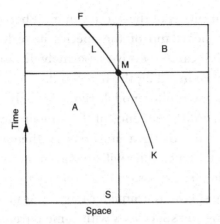

*Figure 10* Discontinuity in space–time.

minima, maxima and saddle-points) form a smooth two-dimensional surface in $X \times P$, as in Figure 9. Call a point of C a *singularity* if it has a horizontal tangent. The deep theorems of catastrophe theory classify the types of singularity that can occur, and there are only two types, fold curves and cusp points, as shown in Figure 9. Moreover they are robust under perturbations of $u$, and so they are universal models for problems with two parameters. Since there are only two types of singularity, Figure 9 illustrates the *most complicated local shape* that C can have. At the same time it is the *simplest global solution* to our particular problem (where simplest means fewest singularities) because the given boundary conditions imply that there must be at least one cusp. Summarising: the cusp catastrophe is the unique simplest universal solution to our problem.

With more parameters there are more types of singularity: Thom calls them *elementary catastrophes* and classifies them in higher dimensions.

## COMPARISON WITH ALLOPATRIC SPECIATION

*Allopatric speciation* is the name given to the following description of evolution. Although it is a commonly-used description it requires, in fact, five somewhat specialised extra hypotheses as follows. There must be (i) a small peripheral subset of the species that (ii) becomes physically isolated from the heartland of the species in order to prevent interbreeding so that it can (iii) evolve separately due to (iv) adverse conditions before (v) re-invading the heartland.

By contrast, our cusp catastrophe model requires only one somewhat mild hypothesis that different ends of the domain favour the evolution of different forms. Using a universal mathematical model we have then deduced that a frontier will appear in the heartland of the species. It is no longer necessary to appeal to allopatric speciation as the main cause of speciation. Similar catastrophe models can be developed by replacing the space axis with some parameter describing the food supply or the types of predator. If different foods or different predators favour the evolution of the different forms, then this can cause bifurcations with sharp divisions within the species.

## MULTIPLE SPECIATION AT PUNCTUATION POINTS

We have yet to explain Figure 2 (right). For this we must go back to the punctuated equilibrium illustrated in Figures 6 and 7 (right) and analyse the catastrophe when $X$ is multidimensional. It suffices to imagine $X$ as two-dimensional. The catastrophe is triggered by the disappearance of a minimum, or more precisely by a gradual change in $u$ causing the coalescence of that minimum with a saddle point. This can be visualised as a pond on a hillside, whose edge is gradually being eroded until it disappears. Generic pictures of the contours of $u$ in $X$ before, at, and after the moment of catastrophe are shown in Figure 11.

Before the catastrophe, the species is in stable equilibrium at the minimum A at the bottom of the pond, and B denotes the saddlepoint. At the moment of catastrophe A coalesces with B at the point C. The contour through C is a cusp whose interior axis points downhill, and therefore the species begins to evolve specifically in that direction. As the species starts going further downhill the direction of

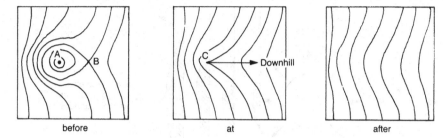

*Figure 11* Contours of u before, at, and after a catastrophe.

evolution is not so clear, and to determine what happens it is necessary to go back to Darwin's original hypotheses (1) and (2). According to (1) the random small variations will cause the species to spread like a wave in all directions, uphill, downhill and along the contours. Meanwhile, the natural selection of hypothesis (2) will prevent it from going uphill, by eliminating the less fit above some contour line. Therefore, the species will begin to travel like a solitary wave rolling downhill at the same time as elongating itself along the contours. The wave-fronts will be curves radiating out from C determined by (1), and the wave-backs will be the contour lines determined by (2). Figure 12 illustrates the artificial special case of equal speeds of variation in all directions and parallel contours. At any given time the species will occupy the sector of a circle as illustrated by the shaded areas. After a while the radius will increase and the species will approximate to a strip between two neighbouring contours.

Darwin makes some pertinent comments in *The Origin of Species* in his initial chapter on the breeding of domestic species. A wild breed is stable under natural selection, and will remain so under domestic selection until the breeder manages to reach the catastrophic point. Darwin then says 'when the organisation has once begun to vary, it generally continues to vary for many generations', corresponding to our wave rolling downhill. Once variation has started 'the whole organisation seems to have become plastic', corresponding to our wave spreading along the contour lines. Furthermore, recall that $X$ is really $n$-dimensional for some very large number $n$, and so the contours are no longer one-dimensional but $(n-1)$ dimensional; hence the

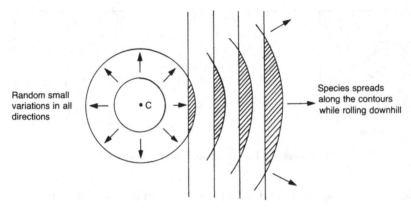

Random small variations in all directions

Species spreads along the contours while rolling downhill

*Figure 12* Catastrophic wave.

species will begin to fan out continuously in every direction but one.

Eventually, the species will finish up distributed amongst a number of minima at the bottom of the $u$-landscape. Individuals in most of these minima are likely to be eliminated by those in fitter minima. The few surviving minima will represent the stable descendant species $A_1, A_2 \ldots A_n$ as in Figure 2 (right). Hence multiple speciation.

Summarising: if there is only one parameter, namely time, then there is only one type of singularity, namely the fold point where a minimum and a saddle-point coalesce, triggering a catastrophe, with rapid evolution and multiple speciation. The result will be recorded as a punctuation point in the fossil record as in Figure 2 (right). This explains why *multiple speciation rather than bifurcation occurs at those punctuation points where there are descendant species*.

We now explain why speciation and canalisation into the new species are likely to occur in mid-evolution, before those species eventually hit their minima and stabilise.

Going back to Figure 12, in general there will be severe problems in measuring the rates of variation and natural selection in terms of measurements of the phenotype. We would not expect the speed of variation necessarily to be the same in all directions, nor would the contours of $u$ necessarily be parallel. For example, consider what will happen if those contours develop some curvature as in Figure 13. The straight dotted lines represent the wave-fronts determined by

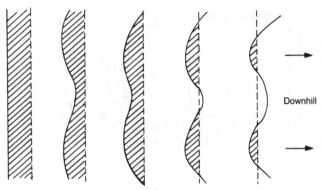

*Figure 13* Speciation and canalisation.

random small variations (1), and the curved continuous lines represent the wave-backs along the contours of $u$, determined by natural
selection (2).

The effect of the curvature of the contours is firstly to separate the
species into disconnected pieces in the valleys between the ridges
(speciation) and secondly to act like a lens focusing each piece into a
small droplet rolling down the middle of the valley (canalisation). In
fact, the effect is enhanced if account is taken of the fact that the wave-
fronts tend to go slightly faster down the steeper slopes of $u$.

Notice that the curvature of the contours below the initial catastrophe point C in Figure 11 (centre) also has the effect of focusing and
emphasising the initial direction of the evolution, before it begins to
fan out along the contours.

Figure 14 shows the path of the species in form–time space, with
the resulting trace in the fossil record. In the expanded non-uniform
time scale of the top picture the rapid evolution begins as a continuous variation of individuals within the species still capable of
interbreeding with one another, and then as the variants diverge they
gradually begin to lose that capability and become separate sub-
species. In the fossil record, however, all this subtlety of variation,
speciation and canalisation is lost in the collapse of the short period
of rapid evolution to a single punctuation point on the uniform time
scale of the bottom picture.

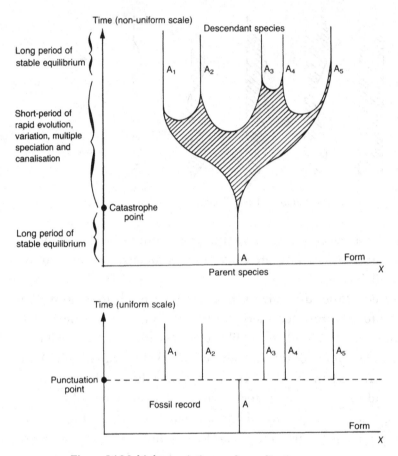

*Figure 14* Multiple speciation and canalisation.

## REMARKS ABOUT DYNAMICS

The familiar concept of the evolutionary tree as a one-dimensional graph has the disadvantage that it contains no dynamics, and therefore offers no opportunity for causal explanation; it is also misleading in suggesting that bifurcations are generic. Modelling evolution with an ordinary differential equation is no good because it cannot explain the central phenomenon of speciation. I do not know whether it would be possible to use a partial differential equation; the difficulty lies in modelling natural selection, because a differential equation is essentially a local concept whereas a contour line is global. It might seem desirable to try and incorporate both processes of (1) random small variations and (2) natural selection into a single equation, but in fact keeping them mathematically separate has helped to clarify our model. Sometimes it can be useful to combine them, as for instance when studying the speed and cross-sectional shape of an evolutionary wave as it rolls downhill.

## CONCLUSIONS

We have used a catastrophe model to show that Darwin's local continuous hypotheses of random small variations and natural selection imply the global discontinuities of

(i)   punctuated equilibria;
(ii)  speciation and canalisation;
(iii) multiple speciation at punctuation points;
(iv)  abrupt frontiers;
(v)   the creation of frontiers; and
(vi)  invasions of territory.

We have also shown that bifurcation is no more generic than multiple speciation.

## FURTHER READING

Darwin, C., *The Origin of Species by Means of Natural Selection*, London: John Murray, 1859. Reprinted, Harmondsworth: Penguin Books, 1984.
Gould, Stephen Jay, *Ever Since Darwin*, New York: W. W. Norton, 1977.
Gould, Stephen Jay, *The Panda's Thumb: More Reflections in Natural History*, New York: W. W. Norton, 1980.
Thom, R., *Structural Stability and Morphogenesis*, New York: Benjamin, 1972.
Zeeman, E. C., *Catastrophe Theory: Selected Papers 1972–1977*, Reading, Mass.: Addison Wesley, 1977.

# 5

# Earthquakes

*CLAUDIO VITA-FINZI*

When in 1755 Dr Pangloss witnessed the destruction of Lisbon by an earthquake that killed 70 000 of the inhabitants, he pointed out to Candide that the event was a manifestation of the rightness of things: if there was a volcano at Lisbon it could not be anywhere else. Without going all the way with Pangloss in the matter of optimism (and none of the way as regards the presence of a volcano in SW Portugal) we may concede that the professor of metaphysico-theologo-cosmologo-nigology had a point. Earthquakes have their place in the economy of Nature, and it is no use wishing that the excess energy responsible for the shaking could be released in some harmless and perhaps more entertaining manner.

Even so, it is difficult to view them dispassionately as mere scientific curiosities. A century ago Charles Lyell, whose *Principles of Geology* spelled the rout of catastrophism, pressed earthquakes into service by arguing that, in the long run, their effects were eminently beneficial. For if there were no earthquakes to counteract the levelling effects of rivers and waves, there would soon be no dry land. But why – lamented Lyell – should the working of this process be attended with so much evil?

Each year the Earth experiences an estimated 100 000 earthquakes sufficiently strong to be felt by a suitably located and attentive observer. One per cent of these will cause damage; a few of the large ones are strategically placed to wreak havoc. As recently as 1976 the Tangshan earthquake wiped out 750 000 people. The 1896 Honshu earthquake produced sea waves or tsunamis 24 m high that left 26 000 drowned. The 1971 San Fernando earthquake did $550 000 000 of damage, the 1989 San Francisco earthquake an estimated 50 billion.

Evidently, before the evil can be countered or at least evaded it must be understood. Earthquakes have long served as a major stimulus for research into the workings of the Earth. Ironically, though by no means surprisingly, the bulk of what is known about the Earth's interior and the processes that mould its physiognomy has been revealed by earthquakes. There is a parallel with the contribution to the study of brain function made by head injuries, a parallel which becomes painful when we recognise how many of the seismic injuries are self-inflicted or owed to human folly (Figure 1).

## EARTHQUAKES AS PROBES

An earthquake is a shock that affects a small part of the Earth and may therefore be viewed as a point or focus. The point vertically above the focus is called the epicentre. The waves that radiate from the focus may be detected by living beings, natural and artificial structures, and instruments. All have something to tell the seismologist, and attempts are constantly being made to squeeze yet more information out of the records obtained by observers and seismometers.

What Bruce A. Bolt has called the seismological age dates from the turn of the century when the development of the seismograph began to reveal, in the manner of X-rays, the 'architecture' of the interior of the earth. A notable landmark is the paper by R. D. Oldham dating from 1906 which demonstrated that variations in the pattern of waves from an earthquake recorded at different locations revealed the presence of a large core (Figure 2).

Not long afterwards, the Yugoslav A. Mohorovičić derived from earthquake records evidence for another discontinuity, this time at a depth of 50 km. Now called the Moho and known to range in depth from 5 to 60 km, it is taken to represent the boundary between the

*Figure 1* Aftermath of the 1980 Irpinia earthquake, southern Italy.

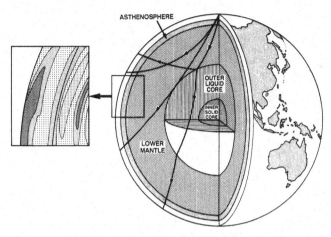

*Figure 2* The Earth's interior. Some of the many possible paths followed by earthquake waves are shown. The inset illustrates the kind of pattern yielded by seismic tomography.

crust and the underlying mantle. In 1914, Beno Gutenberg revised Oldham's estimate for the radius of the core to 2900 km, close to current estimates. In short, the crucial divisions of the Earth's interior were recognised in the space of less than a decade. It took another 20 years for Inge Lehmann, a Danish seismologist, to find evidence for a solid inner core.

The layered scheme in Figure 2 is spuriously neat. The boundaries are inferred from the seismic records. There is a good deal of evidence to suggest that there are intermediate zones between the inner and outer cores and between core and mantle. The lithosphere consists of oceanic and continental crust over the upper part of the upper mantle. Where to draw its boundary with the asthenosphere beneath is a matter of definition. Again, the discontinuity between the upper mantle and the lower mantle at 670 km is thought by some to represent a chemical boundary that effectively insulates the two zones, and by others a mineralogical phase change across which material can migrate.

Moreover, the layers are not laterally homogeneous. For a start, if we are right in thinking that the global machine is driven by heat and that some kind of convective circulation does the trick, there are

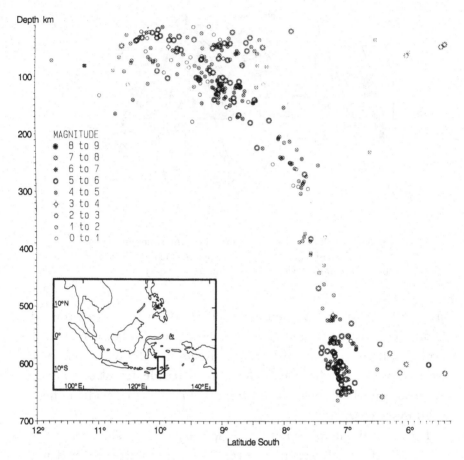

*Figure 3* Downgoing slab defined by earthquakes in part of Indonesia. The period covered is 1964–90 and the events plotted were reported by at least 10 observing stations.

bound to be zones of upwelling and others where material sinks. Then there are places where especially hot zones in the mantle give off 'plumes' which may punch their way to the surface as zones of volcanic activity. Elsewhere, there are downgoing slabs of sea-floor (Figure 3). And everywhere there is the inhomogeneity that characterises any portion of the Earth by virtue of its complex history. Were it not so there would be no volcanoes, mountain-building – or earthquakes.

And here too earthquakes are the source of our knowledge.

Much has already been achieved, even though there are only 100 observatories to monitor the 500 000 000 km² of the Earth's surface. The data amassed by the Worldwide Standardised Seismographic Network (WWSSN) set up in the 1970s, and processed by computers in the light of theoretical and experimental advances, began to reveal by the early 1980s the pattern of lateral differences in density and in temperature (Figure 2).

The data of global seismic imaging – sometimes called seismic tomography by analogy with X-ray tomography – are derived from seismograms in their entirety or from a selected component of the seismogram such as the arrival time of P (primary) waves, which travel through the Earth compressing and dilating the rocks they traverse in the manner of sound waves, or S (secondary or shear waves), where oscillation is at right angles across the path of the wave. (L-waves, the slowest, travel at the Earth's surface.)

## MAGNITUDE

The earliest measures of earthquake force were perhaps inevitably based on damage and on how the earthquake was felt. The first formal scale (Rossi–Forel) distinguished 10 levels of intensity. Plotting intensities on this basis allowed the origin of the earthquake to be plotted and gave some impression of its violence. The Mercalli scale, proposed in 1902, employs 12 levels (I–XII). It remains valuable as it permits a wide range of sources to be pressed into service. It also allows modern earthquakes to be compared with events in the preceding 90 years.

Nevertheless, observers at rest and swinging chandeliers are not always to hand, and even if they are, comparison between different localities is difficult and quantitative analysis problematic. Reliance on standard instruments is the answer. Magnitude was originally defined in 1935 by C. F. Richter for shallow shocks in California as the logarithm of the maximum trace amplitude, expressed in thousandths of a millimetre, registered by a standard short-period seismometer at a distance of 100 km from the earthquake epicentre.

The result is distinguished from other magnitudes by the subscript L (for local). Different calculations are required for deep or distant earthquakes. Surface wave magnitude ($M_s$) tends to be used for the former and body wave magnitude ($m_b$) for the latter. All the formulae are empirical and the uncertainty is likely to be 0.2–0.3.

The $M_s$ scale becomes unworkable with magnitudes above eight as it ignores gross differences in fault size and displacement. Here, seismologists now turn to the $M_w$ scale, which takes into account the seismic moment, a measure derived from the amount of the fault displacement and the rigidity of the rocks at issue. On this basis, the 1960 Chile earthquake, with $M_s = 8.3$, has $M_w = 9.5$, whereas the 1906 San Francisco earthquake with a very similar $M_s$ (8.25) has $M_w = 7.9$.

Richter had in fact made several attempts to render the measurement of magnitude more generally useful by converting it to the energy released by the earthquake. On the basis of the formula he favoured in 1953, the energy of an earthquake of $M_L = 6$ is 40 times that of one of $M_L = 5$ and 1600 times the energy of an earthquake of $M_L = 4$.

## PLANES

Besides the size of the event and its location, the seismograph records reveal how the fault moved. If one accepts that the earthquake stems from fault movement at the focus and that the focus can be viewed as a small sphere, portions of the sphere can be recognised where fault slip leads to compression and others where it leads to its converse, dilation. If motion was strike-slip, displacement occurred in opposite directions along a fault line parallel to the horizontal. The classic example of strike-slip displacement is the San Andreas Fault.

In dip-slip faulting, motion is along a plane dipping or sloping downwards (Figure 4). When the rocks on opposite sides of the plane move further apart, and one of the blocks slides down the slope, faulting is said to be normal. In reverse faulting there is compression and one of the blocks is pushed up the fault plane. There is of course no reason why the earthquake-generating movement should have been

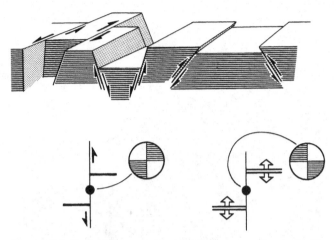

*Figure 4* Major fault types: strike-slip (left), normal (centre) and reverse (right).

wholly of one type. For instance, a dip-slip fault may have a strike-slip component.

The alignment will evidently be influenced by the orientation of existing fractures, as stresses tend to exploit lines of weakness if these are at all conveniently placed. This is one important reason why earthquakes tend to recur in the same localities.

A surface break may make it perfectly obvious how the fault moved during the earthquake (Figures 5 and 6). More often than not the fault does not reach the surface or, if it does, it will be obscured by vegetation, soil or buildings and roads. The seismologist then calculates a focal-plane (or fault-plane) solution based on earthquake records.

The crucial step is to determine the polarity of the first arrival of P-waves at the recording station, that is, whether the first seismic wave recorded represented a push or a pull. The records of the slower S-waves, which are more difficult to read, may also need to be analysed. The work has been greatly speeded up by technical improvements and is now routinely done by computers for key events. The data are generally derived from the global network. Where the solutions are for aftershocks the information may be obtained from portable instruments set up around the main focus and operated for a few days precisely for the purpose.

*Figure 5* Normal faults on the coast of Iranian Makran.

The next stage is to work out where a wave leaving the earthquake focus would cross a small imaginary sphere surrounding the focus on its way to the seismograph in question. (The route will not be a straight line as velocity varies with depth, local geological conditions and so forth.) The polarity recorded for that earthquake at that station is plotted on a grid, usually one that projects the lower half of the sphere onto a flat surface. A good distribution of data points will reveal a set of zones characterised by compression or dilation, which can be bounded by arcs. In Figure 7, which shows the pattern produced by the main shock of the El Asnam earthquake in Algeria in

*Figure 6* Train derailed as it crossed a fault scarp created during the 1980 El Asnam earthquake in Algeria.

October 1980, the pattern of compressional first motions indicates reverse faulting. The pattern would have been reversed for normal faulting. One of the arcs represents the fault plane; which one is a matter of judgment, and here the field evidence may prove of value.

Synthetic seismograms derived from the postulated local geological structure and rock type are also shown above the seismograms that were recorded at selected stations during the first minute or so. They are calculated in order to establish whether the assumptions about local rock type and structure were reasonable. They also help to work out the orientation of the fault plane when it remains ambiguous.

## PLATES

The global distribution of earthquakes was of course one of the keys to the recognition that the Earth's surface could be subdivided into a relatively small number of discrete units or plates (Figure 8). The linear arrangement was already apparent by 1858, when Robert Mal-

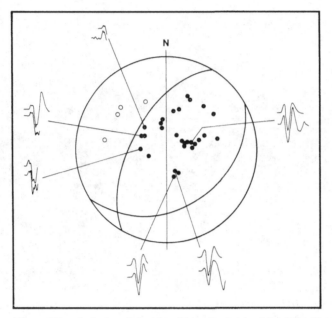

*Figure 7* Focal-plane solution for the main shock of the 1980 El Asnam earthquake in Algeria (10 October 1980). Filled circles: compressional arrivals; open circles: dilatational arrivals. The six synthetic seismograms are above the observed seismograms.

let drew a world map based on historical records. By about 1954, Hugo Benioff had shown that at ocean trenches there is commonly a zone of earthquake foci which dips landwards, often at about 45°, to depths that may exceed 700 km (Figure 3). These two seismological facts are readily explained by subduction, the process by which ocean floors are conveyed beneath the continents and ultimately into the asthenosphere by sea-floor spreading.

The seismicity of the Benioff zones reflects the stresses that afflict the downgoing slabs of lithosphere. The mid-ocean ridges from which the spreading originates are also seismic. The ridges are characterised by offsets. J. Tuzo Wilson proposed that some of the offsets represented a novel kind of fracture, which he called a transform, in which the opposite sides move in a direction contrary to that that one might deduce from the geometry of the offsets because of continued spreading away from the median ridge. Focal-plane solutions of

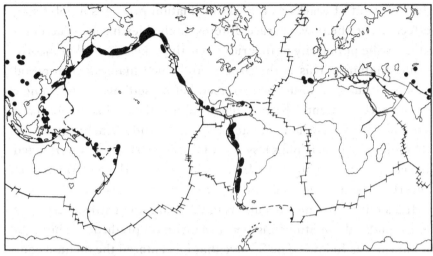

*Figure 8* Main lithospheric plates and location of earthquakes of
$M_s = 8.0$ or greater for 1904–85. The plates are bounded by
divergent (lines) or convergent (toothed lines) boundaries. The
light transverse lines are transform faults, the 'blobs' are earthquakes and rupture zones.

earthquakes on the offsets demonstrated that motion was indeed as
Wilson had proposed.

Some transform faults are sizeable and come on land. The best
known is the San Andreas Fault of the western USA. The Jordan Rift
Valley is another transform along which two major plates slide
spasmodically. Nevertheless, intraplate earthquakes can occur,
sometimes with great violence, witness the New Madrid earthquakes
of Missouri (1811–12), which exceeded X on the Modified Mercalli
scale and were felt on the coast. There is some concern that the seduc-
tive elegance of the plate model could lead to dangerous complacency
over the seismic hazard away from subduction zones or major transform faults, but the fact remains that the subduction zones account
for the bulk of the world's seismicity.

## SPLITTING AND SWELLING

Like variations over space, variations over time reward careful
analysis. What has been called shear-wave splitting occurs when the

development of cracks as a result of dilation produces a polarising effect on the shear-wave trace. Interest in the idea has been accentuated by the popularity of theories of earthquake origins that incorporate dilatancy. This is the process by which materials composed of loose grains expand when stressed before settling down to a new stable arrangement. The best illustration is the dull zone observed around your foot when you step on a wet sandy beach: this is the dilatant area, which looks less shiny than the rest of the waterlogged beach because, having dilated, it is no longer waterlogged. But it soon reverts to being shiny, as water pours into the voids.

If a similar sequence results from the build-up of stress before an earthquake, the splitting might yield a means of prediction. The argument runs as follows. The S-wave that has entered the dilatant zone splits into two phases that travel at different speeds. The delay between their arrivals is proportional to the length of the path through the affected rocks and to the extent of dilation as the cracks are aligned preferentially in one direction. The effect is much heightened by the entry of water into the cracks because S-waves are not transmitted by water.

One might expect a corresponding set of changes in the topography of the site. The classic example is the Niigata earthquake of 1964. Between 1898 and 1955, careful measurements showed that the coast was rising. There ensued 5 cm of rapid uplift, a period of stability lasting a further 5 years, and then the earthquake. Many suspect sites have since been fitted with tilt meters, tide gauges and other devices in the hope of detecting similar premonitory movements. Satellite techniques and in particular GPS (Global Positioning Systems) offer scope for rapid resurvey of extensive areas and are likely to be deployed in the highly seismic islands of Japan.

The countless reports of strange animal behaviour before earthquakes could have a related explanation. They include those by Capt. Fitzroy of the *Beagle* at Concepcion and the reported departure of weasels and millipedes from the doomed city of Helice in 373 BC. The explanations include vibrations, electrical fields, noise and charged particles. The hope remains of identifying the process so that it can be

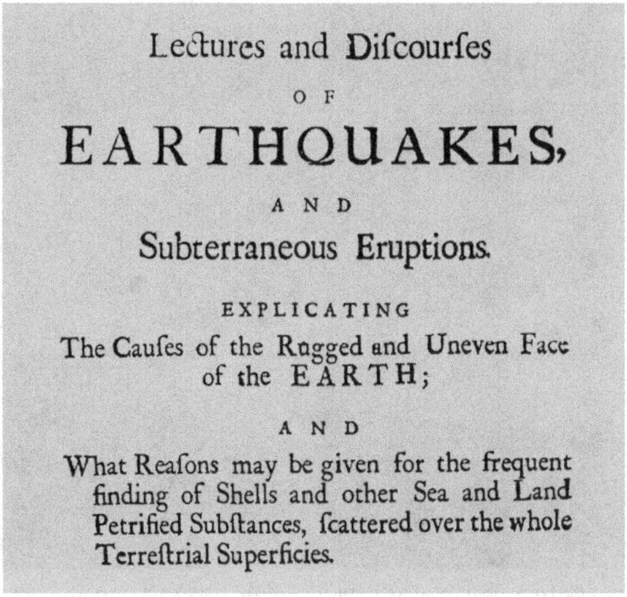

Lectures and Difcourfes

O F

# EARTHQUAKES,

A N D

Subterraneous Eruptions.

EXPLICATING

The Caufes of the Rugged and Uneven Face
of the EARTH;

A N D

What Reafons may be given for the frequent
finding of Shells and other Sea and Land
Petrified Subftances, fcattered over the whole
Terreftrial Superficies.

*Figure 9* Part of the Frontispiece of *A Discourse of Earthquakes*
by Robert Hooke (1688).

exploited as advance warning by other means, just as the death of
canaries is no longer required for the detection of explosive gases in
mines.

## EARTHQUAKES AND MOUNTAINS

According to Lyell, Hooke had been the first (in 1688) to demonstrate
the connection between geological phenomena and earthquakes.
Before that 'the narrative of the historian was almost exclusively confined to the number of human beings who perished, the number of
cities laid in ruins, the value of property destroyed' and the like. In his
*Lectures and Discourses of Earthquakes* (Figure 9) Hooke ascribed to
earthquakes both uplift and subsidence (Figure 10), and under the
first heading he included the elevation of plains, tracts of sea floor and
mountains.

Lyell devoted three chapters of later editions of his *Principles* to the
permanent changes associated with earthquakes. In assessing the
effects of a single earthquake – that which affected Chile in 1822 –

*Figure 10* Aerial view of fields in south-west Chile submerged
by subsidence during the 1960 earthquake.

Lyell expressed the volumes at issue in terms of the great Pyramid. Today we might find it more helpful to compare the energy released with tons of TNT or the number of nuclear bombs; the sense of awe is undimmed.

In sections dealing with the later 1835 Chilean earthquake, Lyell drew on the work of Charles Darwin, who had reported coseismic uplift, in places followed by subsidence, by as much as 10 ft (say 3 m) at Concepcion and on various islands nearby (Figure 11). Darwin was able to conclude that a sequence of such earthquakes could account for a substantial part of the height of the Andes. The well-preserved fossil beaches that Darwin described at various points on the Chilean coast (Figures 12 and 13) support this contention. But the climate of opinion was not favourable to such views. There was a reluctance to abandon the intuitive association between two violent natural agencies, volcanoes and earthquakes. And geological history was based on the assumption that, whereas sea level rose and fell, the lands were stable. Indeed, Eduard Suess, author of the most authoritative text

*Figure 11* Mocha Island, off central Chile. The coastal plain has formed by uplift, partly during earthquakes, and beaches that formed about 5500 years ago are now over 30m above sea level.

*Figure 12* Raised fossil marine beach north of Antofagasta, Chile.

*Figure 13* Shells, some still articulated, from a fossil beach at Caldera, in Chile.

on the subject, *Das Antlitz der Erde*, denied that the Chilean coasts had undergone uplift either during the 1835 earthquake or the 1822 one before it.

The same unwillingness to associate earthquakes with geological change applied to faulting. The elastic rebound theory was put forward by H. F. Reid in 1911. It proposed that strain which has built up by movement in opposite directions across a fault is relieved by sudden slip on the fault. Reid was able to demonstrate the validity of his idea by referring to precise surveys showing how the ground had been deformed on both sides of the San Andreas Fault before and during the earthquake.

Elsewhere, earthquakes are accompanied by the formation of scarps (Figure 14) but the cautious observer is inclined to dismiss such fractures as the effects of landslips. The possibility of folding above buried faults is only now gaining widespread acceptance. The process, manifested at El Asnam (Algeria) in 1980 and Coalinga (California) in 1983, demonstrates another way in which earthquakes can

*Figure 14* Scarp produced by reverse faulting during the El Asnam earthquake of 1980.

promote the growth of mountains. Conversely, folding is now seen as a valuable clue to the location of buried faults capable of generating earthquakes.

## CHRONOLOGIES

All those who have experienced an earthquake, or who design structures intended to withstand earthquakes, wish to know when the last event of a particular size occurred at the same location as the event that prompted the enquiry. Many Londoners who felt the North Wales earthquake of 1990 ($M_L$ = 5.3) and who discovered that an event of comparable magnitude had occurred in the same general area in 1984 were prompted to consider such matters as the frequency of similar events and the odds against anything more violent.

The use of earthquake history as the basis for earthquake prediction has its counterpart in hydrology and meteorology, where such items as the 100-year flood and the incidence of droughts play an important part in evaluating probabilities. Even palaeontology, not usually viewed as a predictive science, now contributes to the field through what it can reveal about patterns of extinction with some kind of recurrence.

Analogies are dangerous and the comparison of earthquake prediction with the work of meteorologists demonstrates it, given the great differences in the data on which they work. But the analogy also offers an escape. Thus, some seismologists are at pains to distinguish between prediction and forecasting, the former being analogous to climatic soothsaying on a grand scale and the latter close to the synoptic approach of the weather forecaster who seeks out local, short-term patterns from the regional picture and from secular trends.

The seismologist may thus identify plate boundaries where earthquakes are likely to arise from subduction or from transform motion (an example of the former being central South America and of the latter central California) much as the climatologist will recognise a belt across NW Europe that is typified by winter depressions. The historical record will then reveal that parts of the key areas have not been affected by significant earthquakes over the period of record, and these 'seismic gaps' will then be viewed with circumspection. The Michoacan earthquake of 1985 fell within one such gap.

Evidently those regions that are not adjacent to active plate boundaries – the intraplate areas that make up the bulk of the Earth's land area – will be disregarded in any such analysis. But most attempts at prediction are on a local scale: if we must salvage the meteorological analogy the scale is that of a single cyclone, and a stationary one at that. Especially when the client is an engineer assessing seismic hazard at a power plant or a town planner revising the criteria used for zoning an urban area, but also in many scientific studies of seismic mechanism, the analysis focuses more often than not on a fault suspected of harbouring seismic tendencies.

Faults are the focus of interest because the success of Reid's elastic model of earthquake genesis has been such that alternative explanations are proposed only for deep earthquakes, that is, in an environment where elasticity and the brittle behaviour responsible for the earthquake itself do not operate. With the realisation that seismo-

*Figure 15* Columns from the Temple of Zeus at Olympia supposedly knocked down by earthquake in the 6th century AD.

genic faults do not always reach the surface, for instance where they are concealed by folds, the range of suspect locations is broadened yet still fault-orientated.

The history of seismicity breaks down into two overlapping phases: historical and instrumental. The difficulty of disentangling earthquakes from other sources of damage or fire is serious enough (Figure 16) without the added problem of assessing the role of earthquakes in the abandonment of a site or the collapse of a culture. In Plato's account of the Atlantis myth, a series of violent earthquakes and floods sank the island empire, as well as destroying prehistoric Athens, in one day and one night. Most of the numerous modern versions of the tale also end in seismic debacle. (Lewis Spence, who had penned one of them, warned Europeans in 1942 that they faced a similar fate unless they mended their ways.) Helice, on the Gulf of

Corinth, was destroyed by a well attested earthquake in 373 BC, but its site has yet to be identified with confidence. What little historical evidence we have is not always immune from doctoring. There is shame in death when one claims the ability to avert it: the figures for Tangshan (Hopei) (1976) were revised down from 700 000 to 242 000 in 1979 by the New China News Agency.

Nevertheless, the search is justified if we are to supplement the brief and patchy instrumental record. Modern Iraq, for example, is poor in earthquakes of $m_b = 5.0$ or greater and it is not immediately clear how far this represents a real contrast with the high seismicity of the Zagros Mountains to the east rather than fortuitous quiescence during the last 30 years. Analysis of the large numbers of clay tablets bearing inscriptions in the cuneiform script promises to answer the question for Mesopotamia during the last three millennia BC. The tablets include royal correspondence and astronomical records, the latter serving to identify the date and time of associated tremors. In some areas, earthquakes emerge as so commonplace as to serve as omens along the lines of 'if an earthquake takes place at dawn the king will suffer a defeat'. Elsewhere, they are rare enough to justify a full description, which is sometimes detailed enough for intensity to be estimated.

Even when all archaeological and literary sources have been fully exploited, geological evidence will be required to plug the gaps and extend the record long enough into the past for a fair assessment of seismic history. How long that should be remains a moot point. In many parts of the USA, a fault is deemed 'capable' (i.e. capable of generating a dangerous earthquake) if it has done so within the preceding 30 000 years. In some studies carried out in New Zealand, the period reviewed for this purpose encompasses the last 500 000 years. Studies now being undertaken in the UK to assess the seismic hazard at sites where nuclear waste is to be stored take into consideration the next one million years in the light of events in the preceding two million.

To extract earthquakes from the geological evidence we use modern examples as guides. In excavated sections the presence of a fault is thus not enough, as it could conceivably have deformed by creeping rather than rupturing instantaneously. Associated sand blows sometimes provide circumstantial evidence of earthquake movement because they reflect sudden ejection of groundwater. Again, raised beaches can form in many different ways. If they are fault-bounded or if they are capped by shell beds that were evidently killed off suddenly, they are legitimate clues to the magnitude and timing of earthquakes.

In all these chronological searches there is often an unspoken long-ing for order. One of the earliest successes was in the correlation between $M_L$ and frequency. In 1949, Beno Gutenberg and Charles Richter showed that there was a very regular relationship for all known shocks, especially if grouped according to depth. For exam-ple, the number of shallow earthquakes (at depths of less than 60 km) increases eightfold if magnitude decreases from 8.0 to 7.0 and again from 7.0 to 6.0. A similar orderly relationship between size and num-ber pattern can be identified on many individual faults.

On the other hand, geology shows that magnitudes greater than seven are underrepresented by such statistical analyses. And the most detailed chronologies fail to reveal anything more than a coarse equivalence in the time interval between successive events of similar size. On the best studied stretch of the San Andreas fault, the inter-vals between the last 10 episodes of faulting range from 44 to 332 years. And different parts of the fault are apparently affected by dif-ferent earthquake magnitudes as well as different periodicities.

The fact that not all movement on an active fault is necessarily manifested in earthquakes helps to explain the imperfection of such correlations. It is also a favoured device for explaining why there is commonly gross mismatch between the energy release to be expected from the movement of interacting plates and that released by earth-quakes. The deficit is especially acute between Eurasia and Arabia, where a mere 10% of the motion indicated by the ocean-floor evidence has been taken up by earthquakes in historical times. But how much of it is due to the vagaries of a defective, brief record rather than to undetected fault creep?

And the future? The dons at the University of Coimbra declared that the sight of a few people ceremoniously burned before a slow fire was an infallible way to prevent earthquakes. Voltaire was pleased to record that on the next day another earthquake caused tremendous havoc. He could not know the prescription would prove effective thereafter. Addison's impudent mountebank sold pills which (as he told the country people) were very good against an earthquake. Current thinking on earthquake control is strongly influenced by the inadvertent promotion of seismicity at several dam sites, both as a glimpse of what could be achieved and how unpredictable the results remain.

## SMALL EARTHQUAKE IN CHILE

In his ruminations on the 1835 Chilean earthquake, Darwin remarked that earthquakes alone were sufficient to destroy the prosperity of a country. To illustrate the case he tried to imagine the effects of violent earthquakes beneath England: Government being unable to collect the taxes and failing to maintain its authority, the hand of violence and rapine would remain uncontrolled.

It is true that the North Wales 1984 earthquake dislocated traffic lights in Dublin. But the weight of evidence is against Darwin. For all their indiscriminate horror, earthquakes seem to have little effect on the flow of history. Ambraseys and Melville describe the aftermath of earthquakes in Iran as typical of many other areas. The local economy is damaged and there ensue population movements, emigration, and increased taxes. Plans for ambitious reconstruction soon wane as funds are found to be inadequate or are misused. The larger sites are not rebuilt and the more active survivors migrate. Yet the effects are almost invariably short-term unless a region was already in decline. On the island of Nias, off west Sumatra, the traditional domestic architecture mimics the galleons that once visited these coasts (Figure 16) and also embodies features that are earthquake resistant. Here the response to repeated tremors was adaptive.

According to Gibbon, the earthquakes of AD 365 'astonished and

*Figure 16* Traditional houses in Nias Island, Indonesia, displaying reinforcement that may have been designed to resist shaking during earthquakes.

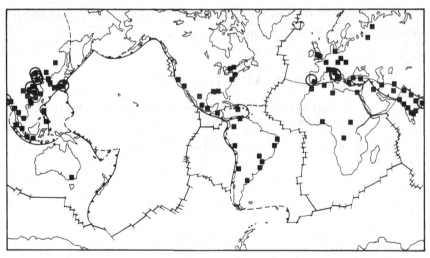

*Figure 17* Location of cities likely to have populations of two million or more in AD 2000.

terrified the subjects of Rome, and their affrighted imagination enlarged the real extent of a momentary evil . . . and their fearful vanity was disposed to confound the symptoms of a declining empire and a sinking world'. Gibbon concludes: 'Man has much more to fear from the passions of his fellow human-beings than from the convulsions of the elements': the institution of great cities, for example, almost realised 'the wish of Caligula that the Roman people had but one neck. Two hundred and fifty thousand persons are said to have perished in the earthquake of Antioch . . . In these disasters the architect becomes the enemy of mankind'. And, as Roger Bilham has shown, many of the fastest growing cities lie close to the great seismic zones (Figure 17).

Gibbon noted the prudence of the inhabitants of Epidaurus, who placed the monk St Hilarion on the beach, for he made the sign of the cross. The mountain-wave stopped, bowed and retreated. The inhabitants of Managua, destroyed by earthquakes 10 times in the last 150 years, refuse to accept defeat. Those of El Asnam, in Algeria, have seen their site destroyed in 1954, when it was called Orléansville, and in 1980, when it had become El Asnam following independence. The city has changed its name to Ech Chelif, doubtless to symbolise its rebirth, but the location remains imprudently the same.

## FURTHER READING

Ambraseys, N. N. and Melville, C. P., *A History of Persian Earthquakes*, Cambridge: Cambridge University Press, 1982.
Bolt, B. A., *Earthquakes. A Primer* (2nd edn.), San Francisco: Freeman, 1980.
Muir Wood, R., *Earthquakes and Volcanoes*, London: Mitchell Beazley, 1986.
Vita-Finzi, C., *Recent Earth Movements*, London: Academic Press, 1986.

# 6

# Storms and cyclones

*NICHOLAS COOK*

The subject of this chapter is the impact of strong winds on buildings and other man-made structures. While strong winds are natural meteorological effects over which we have no control, the strength and therefore the degree of safety inherent in buildings is under our complete control, although this is generally taken for granted by the general public – the phrase 'safe as houses' springs readily to mind. In this context, the concept of a 'catastrophe' is not an absolute. While the loss of an entire city in an extreme wind storm is clearly a catastrophe, the loss of a single house is equally well a catastrophe from the point of view of the owner. I shall discuss the mechanisms producing strong winds and the design process used to predict their effect on buildings, followed by examples of where this process occasionally fails if the structure is unsound or when the design conditions are exceeded.

There are two principal types of man-made structure: firstly, those built by traditional methods which were developed over many years by trial and error without calculations, including most domestic housing; and secondly, engineered structures that have been designed by calculations to withstand the forces they are expected to resist. In modern times these calculations are the responsibility of the structural engineer.

The study of the effects of wind on structures is now given the title 'wind engineering'. As with many other fields, knowledge in wind engineering was gained in steps within various areas of study, each advancing on the strength of the previous steps. Interposed among these advances are the instances of disastrous collapse that initiated or gave added impetus to research. The beginning of research in the United Kingdom stems directly from the loss of the Tay rail bridge in a storm in 1879 just as a passenger train was crossing (the wind loading on the train being a significant factor in the failure), resulting in a great loss of life. This provoked Benjamin Baker, the engineer then in charge of constructing the Forth rail bridge, to make his pioneering experiments on the wind forces on rectangular plates of various sizes at the site of the bridge. The Tay rail bridge failure was a simple 'static' failure, in which the wind loads exceeded the resistance of the bridge structure. Half a century earlier, in 1836, the Brighton chain pier failed in a storm by torsional oscillations, but the mathematical theory, engineering knowledge and experimental facilities necessary to understand this problem did not then exist. The problem was not solved until 1940, when the Tacoma Narrows bridge failed in exactly the same manner, by which time the necessary expertise was available. So the impetus of a disaster is of value only when its cause is capable of being understood; otherwise it is simply dismissed as an Act of God.

The action of wind on structures can be traced through the chain of cause and effect from solar radiation to global circulation, to wind climate, to atmospheric boundary layer, to building structure. This order represents the flow of energy through the chain and the last three links represent the three steps of the design process for engineered structures. Solar radiation is the source of the energy that generates a global circulation through heating the atmosphere more at the equator than at the poles; this was first postulated by Hadley in 1735. The laws of conservation of mass and angular momentum, the latter stemming from the Earth's rotation, produce three main cells of circulation either side of the equator, called Hadley cells. These account for the bands of winds exploited by mariners for centuries: the easterly Trades either side of the equator; the Westerlies at mid-latitudes; and the Polar Easterlies. At the equator where the air from the two tropical cells converges and rises is a region of little wind called the Doldrums. Similarly, at the 'Horse latitudes' where the air from the tropical and temperate cells falls and diverges is another region of little wind. However, the boundary between the temperate and polar cells is different; here warm moist air and cold air meet from opposite directions to form an unstable shear layer called the Polar Front.

Energy is transferred into local weather systems by instabilities in this general circulation, modified by the Earth's rotation and local temperature differences. One element in the formation of these local weather systems is the tendency for flow to turn to the right in the northern hemisphere and to the left in the southern hemisphere, conserving the angular momentum from the Earth's rotation. Coriolis, a gunnery expert in the era of the Napoleonic War, found that cannon balls fired long distances moved to the right as the Earth rotated under the cannon ball in flight, so the effect is named after him. (Coriolis was probably the first to encounter the 'sound barrier' when he found that the range of a cannon increased with the amount of charge up to a limit, but no further.) Another key element is conservation of mass, in that any mass of air that moves must be replaced from elsewhere.

The United Kingdom lies near the top of the temperate cell in the band of Westerlies. Here, the dominant wind-producing weather systems are the frontal depressions formed from instabilities in the Polar Front. A frontal depression forms when a wedge of warm air from the temperate cell pushes upwards and northwards over the heavier cold polar air. Conservation of angular momentum causes this to rotate. The source of energy here is the buoyancy of the wedge of warm air as well as the latent heat of condensation released as the water vapour turns to rain. Once this energy is released, the depression decays and dissipates, but another forms further west along the Polar Front to take its place – a sequence depressingly familiar to all who follow the daily weather forecasts. Typical frontal depressions are very large, over 1000 km in diameter, giving a band of strong wind several hundred kilometres wide, so that about a third of the UK may be affected each time.

Each of the 120 or so meteorological stations across the UK will record the passage of more than 100 depressions each year. The design of a building requires the prediction of the strongest wind in its lifetime. For many years this was done by an extreme-value analysis technique based on the order statistics of the annual maximum values, originally developed around 1930 by Emil Gumbel for predicting maximum river flows for the design of dams in the USA. The maximum wind speed is taken from each year of record and ranked in ascending order. If there were, for example, 45 years of data, the highest value would be assigned a probability of exceedence of 1 in 46, because if the next (46th) year value exceeded the highest value it would be the only one in 46 to do so. Similarly, the second highest would be assigned the probability of exceedence of 2 in 46, and so on. The data lie on a straight line when plotted on special axes, so that a predictive model line can be drawn together with confidence limits.

By using only the highest wind speed value out of 100 storms each year, the standard Gumbel method is very wasteful of data. About ten years ago I developed a method for the analysis of individual storm maxima that gives many more data points, so that the accuracy obtained from nearly 50 years by the standard method can be obtained from only 10 years of data. Further, the analysis can now give the risk by direction, winds from the west in the UK being stronger than from the southeast, which matches the observations of wind damage by direction exactly. Similarly, the wind damage by month matches the monthly analysis of wind speeds, showing buildings to be at their most vulnerable in the winter period. The damage breaks down into two main classes: background damage caused in strong but typical storms each year to buildings that are new and contain design or construction faults and to old buildings that have been allowed to deteriorate; and more widespread damage in individual severe storms coming close to the design conditions. The single storm of 2 January 1976 contributes almost half the total damage in January for the 20 years 1962–81. Similarly, the 'great storm' of 1987 dominates the statistics for October.

Design rules for the UK are implemented in terms of the wind speed with an annual probability of exceedence of 0.02, called the 'once in 50-year wind', which is mapped for the whole country. Most other countries adopt a similar approach. There are two important comments to make here: firstly, that the design wind speed is just the starting point of the design and the loads calculated from this speed are factored up by a 'partial factor for wind' of at least 1.4, which reduces the annual risk to less than 1 in 10 000; secondly, that the concept of the 'once in 50-year wind' as a 'return period' is often misinterpreted badly and needs further explanation.

It is the very name 'return period' that causes the confusion, leading many to believe that, once it has been exceeded, it will not be exceeded again for another 50 years. A more precise name is the 'mean recurrence interval' which correctly implies that the 50-year interval is only an average value when taken over a long period. The risk in any one year is 1 in 50 and, having occurred in one year, the risk in the following year is still the same. A simple model for this process is to imagine a sack containing 50 billiard balls, 49 white and one red. The balls are drawn from the sack at random, one at a time, and replaced. Each draw represents the passage of one year and the drawn ball represents the annual maximum value. The probability of drawing the red ball is 1 in 50, and so represents a year in which the design value is exceeded.

Making this draw 50 times represents a 50-year period. In Figure 1, a notional period of 5000 years is represented, composed of 100 segments each of 50 years. Each dot represents one exceedence of the once in 50-year wind speed and there are 100 such dots, which satisfies the definition of the risk of exceedence, but these are randomly spaced. These occurrences obey the binomial distribution. Thirty-six of the segments are empty, representing 50-year periods where there is no exceedence, i.e. there is a 36% chance of not exceeding the design wind speed in any 50-year period and a complementary 64% chance of having an exceedence. Although an exceedence is more

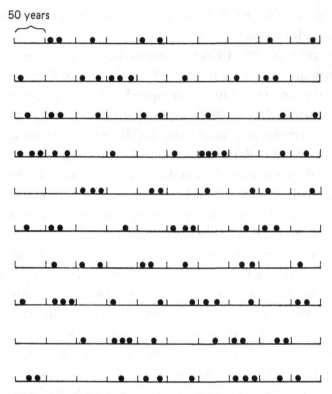

*Figure 1* Recurrence of extremes during a period of 5000 years, each segment representing 50 years. The dots show years in which the 50-year return period event is exceeded or equalled.

likely than not, approximately in the ratio 2:1, it is not inevitable and the chart correctly shows one 150-year period in the 5000 years without an exceedence. As 36 segments are empty, some of the 64 must be multiple occurrences; the chance of two occurrences is 19%, of three is 6% and of four is 1.5%, and the chart correctly shows these. On this basis, the chance of two exceedences in only three years at the same site, for example October 1987 and January 1990, is $1.18 \times 10^{-3}$ or just less than 1 in 1000. In fact, the strong winds in the October 1987 and January 1990 storms affected different parts of southern England and did not exceed the design conditions in the small region of overlap.

The dots on Figure 1 indicate that the once in 50-year wind has been exceeded, but do not indicate by how much it may have been exceeded. About 38 of the 64 will represent exceedences of the 100-year return value, about 18 will represent exceedences of the 250-year return value, and there is a 64% chance that one will exceed the once in 5000-year value. It should now be obvious that, whatever level of wind speed is chosen for a design, there will always be some finite risk of exceedence. The 'partial factor for wind', an additional safety factor of about 1.4 which is applied to the design loads, lowers the effective risk to the completed structure to about 1 in 10 000.

In the tropical cell, the steady Trade winds are interrupted at intervals by tropical cyclones, which are typically a tenth of the size of depressions, but rotate faster, giving higher wind speeds. The strongest winds occur around the eye wall, which is typically 20 km to 30 km in diameter. In this case the energy comes from solar heating of the sea surface to temperatures exceeding 27 °C. As the warmed air rises in the middle of the cyclone it is replaced by more warm air drawn in from the surrounding sea surface by the conservation of mass. This causes the cyclone to spin faster in the middle through conservation of angular momentum in the same way an ice skater increases the speed of a spin by pulling in the arms towards the chest. Unlike the depression, this supply of energy is continuous as long as the cyclone remains over the warm sea. Cyclones form in the summer months for each hemisphere when the sea is warmest. In the western North Pacific in April to December, where they affect China, Hong Kong and Japan, they are called typhoons. In the South Pacific and Indian Oceans from December to March, where they affect Malaysia, the north coast of Australia, India and as far as the African coast, they are called tropical cyclones. In the North Atlantic and Caribbean from August to October, where they affect the Caribbean islands, Mexico and the southern United States, and in the eastern North Pacific, where they affect Mexico and California, they are called hurricanes.

Cyclones rotate clockwise south of the equator and anticlockwise north of the equator because of the direction of the Earth's rotation. If they drift too close to the equator where the component of the Earth's rotation is zero, they lose circulation and collapse. If they drift too far away from the equator, the sea temperature falls below the critical 27 °C threshold, there is insufficient energy supply and they decay. This accounts for the lack of cyclones off the coasts of South America, caused by the cold sea currents from the Antarctic. Once they cross land their energy supply is cut off and they decay quickly, usually within 100 km.

The incidence of cyclones at any one site is much lower than the 100 depressions per year in the UK. Hong Kong has the highest incidence at almost one per year. Regions near the boundaries of the cyclone-prone areas may have incidences of one per decade down to one per century and any individual site may have no recorded instances at all. The analysis of cyclone risk for design must be made in two stages: firstly the analysis of wind speeds of all cyclones in a region by similar methods to those described for depressions, followed by the analysis of cyclone tracks to determine the risk of occurrence at any given site. As cyclones get rarer towards the edges of their ranges, the decision must be made whether or not to design for them. There are no half measures: either a building is built strong enough to withstand a cyclone or it is not, in which case the difference in cost is saved. If the building were to be built with strength between these two cases, it would still fail if a cyclone occurred and the extra strength would be wasted if it did not occur.

The design processes for buildings are intended to cope with these two major sources of strong winds. However, in the small-wind regions between the bands of cyclones and depressions, tornadoes are the principal source of strong winds. Tornadoes are not covered by the design process and any building in their direct path is likely to suffer severe damage. They are formed by local convective instabilities and are often associated with thunderstorms. Tornadoes produce very strong winds in a tight vortex only tens of metres in diameter combined with a large drop in pressure in the core. In the USA they are the most common around the Oklahoma/Kansas border at about 36°N, but they affect the whole of North America. They are also quite common in the UK, but here they are smaller and less intense.

Tornado damage accounts for about a twentieth of the observed wind damage in the UK.

The wind climate can be separated from the following elements of the wind loading chain and given separate independent treatment. This is justified by reference to the total spectrum of wind fluctuations compiled by van der Hoven from measurements at Brookhaven, USA in 1957 and reproduced independently several times since then. This spectrum is a breakdown of the wavelengths of the wind in the same way that white light can be broken down into its various colours. It shows two main peaks: one at low frequencies contains the fluctuations of the wind climate, with a centre period of three to four days, representing the frequency of depressions; the other at high frequencies is the turbulence of the atmospheric boundary layer generated by the flow of wind over the rough ground surface. These two peaks are separated by a distinct gap, the 'spectral gap', and it is this separation that allows the wind climate to be assessed independently. However, the range of frequencies in the atmospheric boundary layer overlaps with the turbulence generated by the flow around the buildings themselves, and the two effects interact and cannot be separated. This requires them to be assessed together.

Let us first consider the atmospheric boundary layer without the presence of any building. The energy of the wind is dissipated by friction with the rough ground surface, creating a layer of turbulent flow over 1 km deep at the boundary with the ground – this is the atmospheric boundary layer. In it the wind speed increases from zero at the ground surface to the full wind speed at the top of the layer and this is called the 'wind speed profile'. Buildings are elements in the rough ground surface; they are totally immersed in the atmospheric boundary layer and experience the energy flow as imposed loads. Changes of ground roughness, such as the edge of a city, change the characteristics of the wind speed profile and the turbulence. Topographic features such as hills and escarpments accelerate the flow, increasing the mean wind speed at the surface by up to a factor of 1.6, and this must be accounted for in the design process.

Recently, accurate models of the atmospheric boundary layer characteristics have been developed. To exploit these models, the Building Research Establishment (BRE) has recently compiled a database of the ground roughness of every 1 km square of the UK. This database provides the input data for a computer programme called 'STRONGBLOW' which takes the roughness changes radially out from the site, together with any topography, from which it is able to calculate the characteristics of the wind incident at the site.

Moving on to the structure itself, owing to the interaction between the atmospheric boundary layer and the structure, it is essential that the atmospheric boundary layer is represented in any studies of the building. A great deal of early research was conducted in low-turbulence uniform flow aeronautical wind tunnels which gave erroneous or misleading results that confused the understanding of the problem for some years. It was Martin Jensen, a Danish engineer, who did the pioneering work in scaled models of the atmospheric boundary layer in the early 1950s. Nowadays, all serious research at model scale is done in wind tunnels specifically designed to represent the atmospheric boundary layer at the same scale as the model building. Basic building shapes are used to build up information on the pressures on common forms of buildings. Small holes in the outer skin duct the pressure through tubes to a transducer inside the model. In this way, general design guidance can be compiled that the designer can use for most common buildings.

Unusual building shapes require individual testing in a boundary-layer wind tunnel. These are special wind tunnels that reproduce the turbulent structure of the atmospheric boundary layer by having a long rough floor over which a scaled version of the atmospheric boundary layer is grown. Different forms of roughness represent different types of terrain: for example, a layer of gravel will give a good representation of open-country terrain at about 1:5000 scale; while plastic coffee cups of the type used in vending machines produce a good representation of town terrain at about 1:300 scale when glued upside down to boards on the floor. In a new boundary-layer wind tunnel currently under construction at BRE, the ground roughness is formed by flat plates, which slide out of the floor and may be adjusted to any height. This boundary-layer growth is accelerated by other devices, usually a turbulence grid and a serrated wall, placed upwind of the roughness so that the wind tunnel appears to the flow to be much longer. Taken together, the effect of these devices is to mimic the effect of the rough ground surface so that even the individual gusts of wind are correctly represented in size and intensity. Neighbouring buildings are also represented accurately so that their individual effects on the building are reproduced.

Even if an adequate design assessment can be made from a code of practice or other standard design guidance, wind tunnel tests are often performed on prestigious developments to obtain more detailed and accurate data than a code can supply. The wind tunnel is most useful when the building is of an unusual shape that could not reasonably be expected to be found in a standard code of practice.

No matter how well the design wind loads are predicted, all effort is for nothing if the building is not correctly designed or constructed to meet them. I sometimes wonder whether the story of the Three Little Pigs and the Big Bad Wolf should be required reading for all builders. In the general loading case for a typical dwelling house there is positive pressure on the windward wall, and suctions on the side and rear walls. When the pitch of the roof is less than about 30°, there is suction on both upwind and downwind slopes. However, when the pitch is greater than 30°, the upwind slope will have positive pressure. The maximum positive pressure is the kinetic pressure of the wind when brought to rest, but the maximum suction is theoretically unlimited. In practice, the maximum suctions can be higher than twice the maximum pressures, leading to suction being the most common mode of failure.

In the UK, falling masonry is by far the major cause of death, toppling chimneys being the most common cause, followed by gable-end walls, then masonry boundary walls. Although cladding failures can be spectacular, they tend to cause injury rather than death.

The chain of cause and effect passes through the building from the pressures and suctions applied to the outer skin, through the structure to its foundations. Failure occurs at the weakest link. So if the outer cladding is not securely attached it will be stripped off. If it is secure, but the roof structure is insufficiently strong, part of the roof may be lost. If the roof structure is not tied to the walls the whole roof is lost, in which case the roof may cause damage to adjacent buildings. All these forms of failure occur in wind speeds below the design value to the most exposed buildings with design or construction faults.

Engineered structures fare much better than those built to traditional rules. Most failures are to the cladding, to traditional brick used as cladding, or curtain walls. Total collapse of engineered structures is rare and tends to happen to structures with little redundancy. Cooling towers are pure shell structures and the most famous collapse at Ferrybridge occurred to towers in the downwind row, caused by the wind accelerating between the upwind towers. The more recent collapse at Fiddler's Ferry occurred to a tower that had deteriorated and was cracked. Another form without much redundancy is the portal frame building.

The storm of October 1987 came as a shock to southeast England. Although the west coast of Scotland experiences these wind speeds every couple of years, the return period for some locations in Kent approached 250 years. Even so, with wind loads only 16% higher than the design case against a minimum 40% safety factor, there should have been no significant damage. As in previous storms, temporary structures fared poorly, including temporary school classrooms; roofs were damaged in all the stages described earlier, including complete removal; chimneys toppled and masonry gable ends failed, especially in older buildings. A feature of this particular storm was the extent of tree loss and damage to buildings from the falling trees. Apart from this, however, damage was no different than on previous occasions. Anyone wanting further information on UK damage will find the statistics of the 1987 storm in the BRE report of the storm; there is also an illustrated review of general UK damage to non-engineered structures, and both items are listed at the end of this chapter.

Let us now move on to damage caused by tropical cyclones. Tropical cyclone Althea crossed the east coast of Queensland, Australia, 48 km north of Townsville on the morning of 24 December 1971. This was not a severe cyclone, the maximum gust speed recorded in Townsville of 53 m/sec being comparable with the design wind speed for Scotland. Of a survey of 6000 dwellings, 0.7% were destroyed, 1.7% had major but repairable damage and 13.3% suffered minor damage. The patterns of damage to buildings less than 5 years old were studied in some detail in order to assess current design and construction methods. Most of the damage was to roofs, 56% being structural damage and 29% to roof cladding, with structural damage to walls only 21% of the damage. Only 12% of the damage was attributed to poor workmanship. Building regulations were tightened up after Althea and, in particular, roofs were required to be securely fixed down through the walls to the foundations.

Exactly three years later, tropical cyclone Tracy passed over the city of Darwin in the early hours of Christmas Day 1974. Tracy was a small but intense cyclone that was particularly slow moving, so that the city was subjected to strong winds for about four hours. The maximum gust speed was estimated at about 70 m/sec, resulting in pressures and loads twice that of Althea, exceeding the design loads and safety factor. Traditional housing suffered badly, 53% being completely destroyed and only 6% remaining habitable. In those cases with the improved roof ties, the roofs remained in place, but the houses collapsed by racking failures of the walls. In amongst the devastation of the housing remained engineered structures that were undamaged, or that had only minor impact damage from flying debris.

The performance of apartment blocks was better, with 10% destroyed and 50% with only minor damage or intact. Unlike in Althea, wind-borne debris played a major part in the Tracy damage, the additional wind speed carrying the debris further, striking other buildings and initiating more damage in a chain reaction. This was a common mode of progressive failure: firstly, windows on the windward face were broken by debris, allowing the wind to exert pressure on the inside, adding to the uplift of the roof. Then, the roof of the first room lifted away and the windward wall collapsed. The internal wall to the next room was then the windward wall and, not being designed for the external wind pressure, this collapsed also. Now the wind in the next room lifted the next section of roof, the next wall collapsed, and so on.

Fully engineered structures behaved very well, with only 3% destroyed and 80% intact or with minor damage. Those few engineered structures that did collapse, mostly portal-frame buildings, were later found to have been designed for a gust wind speed of only 40 m/sec, intended for the cyclone-free areas of Australia, so had experienced three times their design loads.

If that was the experience of a severe cyclone in a country with corresponding building regulations, what happens in a minor hurricane where there are no regulations? At midnight on 21–22 September 1989, hurricane Hugo crossed the coast of South Carolina, directly over the city of Charleston. The central eye passed directly over the city. The track was to the northwest, normal to the coastline. Wind speeds were strongest at the eye wall; the wind direction at the front of the eye was northeast with gusts up to 96 mph and at the back of the eye was southwest at the same speed, that is parallel to the coast in either direction, with a calm period in the eye lasting about 30 minutes, which was later described locally as the 'looting time'. To the northeast side of the eye, the translational speed added to the rotational speed to give gusts up to 110 mph.

An elevation of sea level, called a storm surge, accompanies cyclones, caused partly by the decrease in atmospheric pressure in the eye and partly by the winds 'piling up' the sea against the coastline. In the centre of the track, the surge was about 1.2 to 1.8 metres, but on the northeast side it was 2.4 to 3.6 metres. There were two main types of housing along the coastal strip, older 'low-set' houses built directly on the ground and newer 'high-set' houses built 2.4 metres above the ground on piles. There are no building regulations for the area. Householders may build as they wish, but construction is mostly timber frame with some unreinforced masonry. The standards are

*Figure 2* Low-set house at Folly Beach after hurricane Hugo,
21–22 September 1989. Wave damage.

effectively set by the insurance companies. Only the houses on piles
qualify for flood insurance.

Figure 2 shows what was a low-set house at Folly Beach, north of
Charleston, where the wind and surge were strongest. This was probably entirely wave damage. The house had been swept off its foundations, but the roof structure was virtually intact. Immediately
adjacent to this house, directly on the beach, was a two-month old
house on piles. Figure 3 shows the view from the beach. This had
failed due to wind alone, by the progressive mechanism described
earlier, only in this case the side walls had been lost also.

Offshore from Charleston is a chain of barrier islands that are little
more than sand dunes, one or two feet above high tide. Many low-set
houses were destroyed like the earlier example. Others lifted intact
off their foundations and floated away until stopped by trees, as
shown in Figure 4, or collided with other buildings. Of the high-set
houses on piles, roof removal was again the commonest mode of
failure, sometimes aggravated by the extension of the roof to form a

*Figure 3* High-set house at Folly Beach after hurricane Hugo. Wind damage.

canopy over a porch. The combination of suction on the top and pressure on the underside of the canopy led to failure of the roof, followed by progressive failure as in Figure 3.

Other roof failures were a little more puzzling until investigated. Of a row of condominiums, some retained and some lost their roofs, apparently at random. Closer inspection showed that those units without their roofs had had their patio doors blown out of their rails; wind blowing into each unit through the doors had then lifted away the roofs.

There were also extensive racking failures due to insufficient bracing of the framed buildings. In all, the standard of construction in the area is quite poor, with insufficient attention being paid to details. The detailing is the key and it would cost very little extra to rebuild to a standard that would resist the loads experienced. Unfortunately, the general attitude amongst homeowners and builders alike is that the hurricane was unsurvivable, so that the housing is being rebuilt to the previous poor standard and on sites that are patches of sand,

*Figure 4* Low-set house from barrier islands offshore from
Charleston, after hurricane Hugo. It lifted off the foundations
and floated away until stopped by trees.

60 cm above sea level, which may have cost up to $90 000 for a 15-metre square plot.

The only source of strong winds that is truly unsurvivable is the tornado. That is not to say that it is impossible to design against a tornado, just uneconomic; since the tracks are so narrow, the risk to any individual building among the whole is tiny. A typical track made by a tornado in the USA is a strip two blocks wide in which all houses are completely demolished, parked cars removed and lawns excavated to a depth of around 40 cm. Either side of this strip, houses have their roofs removed, more on the right-hand side where the rotational and translation speeds are compounded. On either side there will be bands of minor roof damage, tile removal and window breakage, and elsewhere no damage at all. Sensible practice in tornado-prone areas is to build an underground storm shelter or a reinforced-concrete windowless room, perhaps the bathroom, in which to take refuge.

These tornadoes are sufficiently strong to sweep away pedestrians and cars, and to overturn railway engines. Many deaths and major injuries occur indoors from flying debris. A piece of 'four by two', the standard US timber size for timber framing, will penetrate a house wall. Ordinary playing cards have been found embedded on edge more than an inch into wooden shipboard cladding. Tornadoes also have a strange effect on poultry and other birds: the pressure drop in the core is more than the suction that holds their feathers in place and the result is instant plucked chicken.

This brings me to the end of my chapter. In essence, the key to turn catastrophe into success is good prediction of the wind climate, good design guidance, appropriate building regulations and sound building practice.

### FURTHER READING

Buller, P. S., *The October Gale of 1987*, Watford: Building Research Establishment, 1988.
Buller, P. S., *Gale Damage to Buildings in the UK – An Illustrated Review*, Watford: Building Research Establishment, 1986.

# 7

# Famine in history

*PETER GARNSEY*

'Nothing is more shamelessly demanding
than an empty belly,
which commands attention,
even if the body is heavy with weariness,
the heart laden with grief.'

HOMER (Odysseus)

'And waking early before the dawn was red
I heard my sons, who were with me, in their sleep
Weeping aloud and crying out for bread.'

DANTE (Count Ugolino)

Hunger has always been part and parcel of the human experience – in archaic Greece, late medieval Italy, Tudor England, Stalinist Russia, and present-day Asia, Africa and Latin America.

Indeed, hunger, undernourishment, malnutrition rather than famine, is the scourge of the developing nations today: this is a theme of Roger Whitehead's 'Famine' lecture delivered in an earlier Darwin lecture series. In the Third World, hunger is endemic and universal. In the words of the famine theorist Amartya Sen:

Most often hunger does not take its toll in a dramatic way at all,
with millions dying in a visible way (as happens with famines).
Instead, endemic hunger kills in a more concealed manner.

People suffer from nutritional deficiency and from greater susceptibility to illness and disease. The insufficiency of food, along with the inadequacy of related commodities (such as health services, medical attention, clean water, etc.), enhances both morbidity and mortality. It all happens rather quietly without any clearly visible deaths from hunger. Indeed so quiet can this process be that it is easy to overlook that such a terrible sequence of deprivation, debilitation and decimation is taking place, covering – in different degrees – much of the population of the poorer countries in the world.

## FAMINE AND HUNGER

Endemic hunger – the subject offers multiple opportunities for the historian. Did malnutrition exist on a large scale in the past? Were pre-industrial European, or traditional Mediterranean societies, in this respect 'third worlds'? These are questions historians have hardly begun to ask themselves. This is partly because the primary sources on which we depend have not conceptualised the *state* of endemic hunger, as opposed to the *event*, going hungry. Famine, in contrast, is no novelty to the historian. On the other hand, general treatments of the phenomenon by historians are rare. David Arnold's book is an honourable exception. Historians of famine tend to be well-versed only in their own famines. Thus the subject of famine poses a challenge too.

## FAMINE AS CATASTROPHE

Is famine a catastrophe? The question may seem otiose. First, is not hunger-induced death the most pitiful way to die? Count Ugolino in the Ninth Circle of Hell, again:

> I gnawed at both my hands for misery:
> and they, who thought it was for hunger plain
> and simple, rose at once and said to me:
>
> O Father, it will give us much less pain
> If you will feed on us; your gift at birth
> was this sad flesh, strip it off again.

Multiply the Ugolino family by hundreds of thousands, and there is a disaster too painful to contemplate: the Ukraine in 1933; Bengal in 1943; China in 1959–61.

In famine, then, the burden of human responsibility is heavy. There is commonly a human input in the form of negligence, selfishness, maladministration, ideological blindness or dogmatism, without which there would be no famine. In this respect, famine is more comparable to war than to other human catastrophes.

But there is a problem: the 'frequency' of famine is a standard theme of historical writing. If famine has been a frequent occurrence in history, can it be, by definition, catastrophic? What is the likelihood that catastrophe in the form of famine struck China between 206 BC and AD 1911 on no fewer than 1828 occasions (to cite an often repeated statistic)? Or that catastrophic famine, to quote Fernand Braudel, the eminent historian of early modern Europe and the Mediterranean world, 'recurred so consistently for centuries on end that it became incorporated into man's biological regime and built into his daily life'? Braudel reports, for example, that the city of Florence in around 400 years between 1371–1791 experienced 16 'very good' harvests and 111 years of 'disette'. This word, which should mean 'shortage', is more or less interchangeable with 'famine' in Braudel's discussion. This may point the way to a resolution of the problem. Perhaps famine has been confused with associated or similar phenomena.

There is need, in my view, of a ground-clearing operation, which sets out to clarify the nature of famine, to lay down guidelines as to how to identify it in the records of the past, and to point to trends in the historical development of famine. Such is the purpose of this paper. I do not present detailed case-studies of particular famines, but rather use those already in existence as a basis for generalisation. (I gladly acknowledge my debt at this juncture to all 'faminists', past and present, without whose work this chapter could not have been written, not forgetting the father of them all, the one time Fellow of Jesus College, Cambridge, Thomas Malthus.) I will not be concerned directly with *mentalité*, or attitudes, and with responses to famine, secular and religious. Nor will I deal with the treatment of famine in

*Figure 1* Starving Bedouin, Unas Pyramid Causeway, Sakkara, Egypt. 5th Dynasty, *c.* 2350 BC.

imaginative literature, from *Revelation* to *The Grapes of Wrath*. I make selective use of art. Let this unfortunate Bedouin of around 2500 BC represent the misery of famine sufferers through the ages.

## FAMINE AND SHORTAGE

Famine is not endemic hunger, as I have already implied. Endemic hunger, or chronic malnutrition, is a condition of long-term food deprivation, whereas famine is a particularly acute food emergency. The two phenomena may be closely related. In some settings, for example in pre-industrial Europe and in the modern developing world, the state of endemic hunger is periodically punctuated by episodic famine; while a famine emergency characteristically sinks a society deeper into the mire of chronic undernourishment and poverty. Thus Garcia and Escuderos can talk daringly of famine and hunger together with drought as bringing 'constant catastrophe' upon the contemporary Third World.

Famine is not shortage. Both are hunger-related crises, but of different degrees of seriousness. Historians tend to collapse famine into shortage, or variants such as dearth, scarcity or hunger; the same writers, significantly, write of famine as a frequent occurrence. A consequence is that a qualitative account of famine is hard to derive from their discussions. We cannot say on the basis of Braudel's account how many of the 111 Florentine food crises were catastrophic: how many, in my terms, were famines.

An Egyptian landowner called Hekanakht, who lived around 4000 years ago, points the way forward. In a frosty letter to some discontented dependants, he wrote:

> I have managed to keep you alive until this day. Take heed that you do not fall into anger ... Being half alive is better than dying altogether. One should use the word hunger only in regard to real hunger. They have begun to eat people here.

Following Hekanakht, we should build into our concept of famine a recognition that there are different levels of human suffering and social dislocation; and we should be prepared to give famine a strong definition.

My provisional summary definition of famine is as follows:

> *Famine* is a critical shortage of essential foodstuffs, leading through hunger to a substantially increased mortality rate in a community or region, and involving a collapse of the social, political and moral order.

A subsistence crisis that is less than a famine I call a food shortage, defined thus:

> *Food shortage* is a short-term reduction in the amount of available foodstuffs, as indicated by rising prices, popular discontent and hunger, in the worst cases leading to death by disease or starvation.

I use *food crisis*, or subsistence crisis, not as a synonym of food shortage, but as an umbrella term, encompassing any kind of food emergency.

I submit that famines, thus defined, are, and always have been, rare; they are genuine catastrophes. When historians and commentators say that famines were frequent, they are actually talking about food shortages, or lesser food crises.

To illustrate: let us glance at a food crisis in Florence in 1329, which I would characterise as a shortage rather than a famine. We are given a privileged view of this crisis through the eyes of Giovanni Villani, a contemporary chronicler and government official, and Domenico Lenzi, grain dealer and market official of Orsanmichele, the grain market in the centre of Florence. Lenzi kept a ledger (*Il Libro del Biadaiolo*) recording the prices of grains and legumes in 1320–35, filled out and enlivened with his own observations and reflections. From data he presents it can be seen that in June 1329 the price of wheat was roughly four times higher than in the previous June, and that other seed crops (rye, beans, barley and spelt) rose with wheat (Figure 2).

Lenzi's book is ornamented with full-page miniatures. One shows a harvest scene in times of plenty (Figure 3). An angel is trumpeting the good news of joy (*allegrezza*) and abundance (*abbondanzia*). But there are ominous rumblings coming from the third trumpet. God issues the warning: 'I can take it all away; you had better be grateful'.

The companion miniature (Figure 4) depicts a harvest scene in less happy times. The devil is in the ascendant, having routed the angel and broken his trumpet. God summons the defeated angel: 'Come back to heaven; it is cleaner and purer here'.

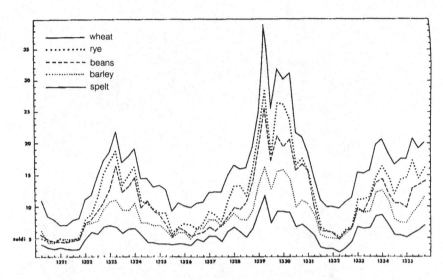

*Figure 2* Cereal and Dry Legume prices, Florence 1320–35.

In a second pair of miniatures, we see the grain market in happy and in unhappy times. In the first, the scene is orderly. The grain bins are full, there is *allegrezza*, singing and dancing below (Figure 5). In the second (Figure 6), there is confusion in both heaven and earth. The angel is in retreat and the trumpets are broken. The dialogue reads:

> *Angel*: I am happy and content in my refuge.
> *God*: Nourish the soul, let the body be punished.
> *Devil (to God)*: I will do as you have permitted me.
>   *(to the crowd)*: Weep, you have reason to. The good is past.
> In hunger and want I shall make you suffer.
> *Crowd*: Grief upon grief. God is abandoning us to the worst of
> fates.

So the miniaturist; Lenzi has the crowd shouting something like this (I paraphrase):

> Those merchants are behind the shortage. We should do away
> with the lot of them and make off with their goods . . . A curse
> on those corrupt politicians for leaving us without food. Let's
> go to the houses of the robbers who are hoarding the grain and
> burn them down, with their owners too, for starving us.

*Figure 3* Harvest Scene, Plenty.

According to Villani, the officials met the risk of disorder by calling in the commune militia. They also had an execution block and axe brought into the piazza, and threatened to amputate the foot or hand of anyone caught stealing.

*Figure 4* Harvest Scene, Shortage.

In the miniature, we see the people battling for such grain as is available, while the commune militia stands close by to protect grain and merchants.

But in the side panel charity is shown in operation. Food is being distributed to the established categories of 'respectable' poor through

Figure 5 Grain-market, Plenty.

the agency of the Confraternity of the Miraculous Madonna of Orsanmichele.

This is a shortage, not a famine. Prices have rocketed, the people are discontented. Food is short, and there is a scramble for residual

*Figure 6* Grain-market, Shortage.

grain. For there is some grain, the barrels are low but not empty, charities are active, and the little people (*popolo minuto*), according to Villani, are receiving rations from the commune. There is even something left for the strangers and rustics, who, having been (according to the fiercely patriotic Lenzi) thrown out of 'perverse tormenter and lunatic' Siena, were received and fed at Florence

*Figure 7* Siena, Expulsion of the Poor.

(outside the walls, however: see Figures 7 and 8). Finally, Villani says nothing about deaths.

In contrast, Ireland in the Great Potato Famine of 1846–50 experienced at least 1 million deaths and an overall rise in the crude death rate (the numbers dying per thousand of the population) of 100%. Or, take the Soviet Union in 1933. The crude death rate for the European provinces of the USSR in 1933 was about double that of the previous year; in one of those provinces, the Ukraine, it tripled; and within the Ukraine in the city of Kharkov it almost quadrupled (see Figure 9).

*Figure 8* Florence, Feeding the Poor.

## THE FOOD CRISIS CONTINUUM

The famine/shortage dichotomy is not a precision tool by which historians can readily differentiate between catastrophic and lesser food crises. The idea of a dichotomy holds out less promise than that of a spectrum or continuum of food crises. Any particular food crisis occupies a place on a continuum leading from mild shortage to disastrous famine.

The food crisis continuum has two main advantages. Firstly, it does not imply the existence of a distinct boundary between famine and

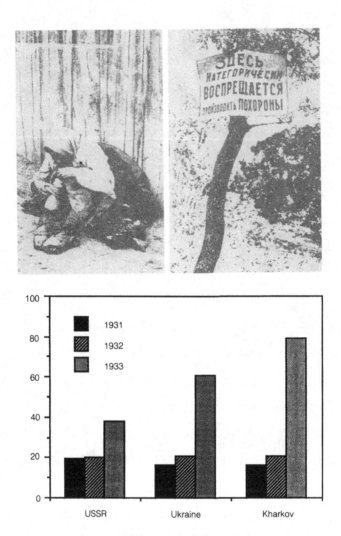

*Figure 9* Above, Scenes from Kharkov, Ukraine, 1933. Left, Waiting for Death. Right, Notice in park 'It is forbidden to bury the dead here'. Rises in crude death rate in USSR, Ukraine and Kharkov, 1933.

shortage. It is not at all obvious how such a boundary could be demarcated.

Secondly, the food crisis spectrum is more compatible with the idea of famine as a dynamic process, in the course of which a population moves in stages from dearth through destitution to considerably increased mortality and social disintegration. Advocates of this view are reacting against what they see as a tendency to focus exclusively on the last stage of famine. My position is not necessarily incompatible with theirs; it is that only those subsistence crises that have evolved as far as that terminal phase are properly termed famines.

A strong definition of famine, one that reserves the word for the most severe food crises – for catastrophes – does create problems in approaching historical societies. It exposes one to the accusation of imposing an academic definition that is at variance with actual historical usage. Documents that survive from the ancient, medieval and early modern worlds use the word famine or an equivalent to describe events that might not qualify under my definition.

This is to broach a problem familiar to historians, namely, the interpretation of primary source materials. Here it is appropriate to refer to a critique of the use of literary sources by modern historians, presented with characteristic forthrightness by Cambridge historian and former Master of Darwin, Moses Finley. His criticisms apply, not just to historians of antiquity, but also to medievalists and early modernists. The gist of his complaint is that we tend to approach historians of past societies as if they possessed the conceptual framework and technical skills with which the historian of today is equipped. The historian in approaching his sources should be fully aware of the differences between his and their intellectual horizons and traditions.

Finley had in mind authors such as Herodotus, Thucydides and Livy. But we often have to deal with annals and chronicles that are much more rudimentary, much less high-genre, than the works of mainline historians such as these – works such as those that furnished the raw material for Sir William Wilde's *Table of Cosmical Phenomena, Epizootics, Famines and Pestilences in Ireland, 900–1850* (see Figure 10). There are considerable obstacles in the way of composing a qualitative account of famine in medieval Ireland on the basis of entries such as these.

The entries lengthen as we approach Wilde's composition date of

1099 'Great dearth of provision in all Ireland.' [*Annals of Ulster*]
'Plunderings, and the evil deeds of war and famine . . .'
[*Annals of Innisfallen*]

1137 'A great scarcity [*tacha*], in the province of Connaught, of which multitudes died.'
[*Annals of Kilronan*]

1153 'A great famine raged in Munster, and it spread all over Ireland, being occasioned by the vehemence of the war.'
[*Annals of Innisfallen*]

*Figure 10* Extracts from Wilde's *Table of Cosmical Phenomena, Epizootics, Famines and Pestilences in Ireland, 900–1850.*

1851, as elaborate eyewitness accounts become available. But they have a rhetoric of their own that is hard to penetrate. I illustrate this time from Scotland. Robert Somers composed his *Letters from the Highlands: After the Great Potato Famine of 1846* for the North British Daily Mail after touring the Scottish Highlands in the autumn of 1847:

> Anyone who witnessed the groups of wretched creatures who crowded into our large towns during last summer and autumn – who knows the want and privation which there awaited them – who saw hundreds of families lying night after night on the cold damp grass of Glasgow Green, or amid the still more pestilential vapours of the wynds and lanes, and who listened to the barking coughs of the infants, as if their little bosoms were about to rend, can require no statistics to satisfy their minds of the fearful destruction of human life occasioned by the ejectment of the peasantry from the parishes in which they were born and had lived, and the property of which should have been made responsible for their sustenance in the day of

> famine. This country was last year the scene of a Massacre of the
> Innocents, which has had no equal since the days of Herod the
> Infanticide.

One can admit the value of Somers' work while recognising its limitations as a historical document. 'A Massacre of the Innocents' 'no equal since the days of Herod the Infanticide' 'anyone who witnessed [it] . . . can require no statistics'. It was precisely the mortality figures, much less impressive in Scotland than in contemporary Ireland, which led Devine to raise the question whether the title famine was merited in this case.

In sum: the literary sources will make most sense to the historian of famine if they are approached with a firm set of criteria for famine, drawn from a wider survey of food crises than any primary sources can control, and arrived at with the aid of a conceptual framework lacking to earlier historians and chroniclers.

In practice, the debate about the meaning and nature of famine is conducted almost exclusively among social scientists and Third World experts. It goes without saying that historians of famine in past societies can ill afford to be ignorant of the course of their lively and ongoing discussions. Does the preceding argument need modification in the light of the debate?

My strong definition of famine appears to be consistent with the usage of Amartya Sen and some other modern famine experts. In a recent paper, Frances d'Souza shows an interesting reluctance to classify as famines the 'food emergencies' in Lesotho in 1983–85 and Mozambique in 1982–85. The title of famine is reserved for the Ethiopian experience (1983–85), for the reason that of the three countries in question, only Ethiopia suffered mass starvation and long-term disruption.

This position must meet the challenge of Alex de Waal. His argument is that famine must be given a broad and flexible meaning, one which (in particular) embraces not only 'famines that kill', but also emergencies that are not markedly destructive of human life, if, that is, one is to produce analyses that are consistent with the perceptions of famine sufferers themselves. And this is an imperative, not a mere option, because, to judge from de Waal's experiences in the Sudan, the judgments of famine sufferers are more accurate, their understanding of their predicament and how to deal with it deeper, than that of visitors from outside, however expert.

De Waal's polemic against the 'disaster visitors', typically the representatives of international agencies and relief organisations, is sustained and effective, and appears to undercut my advice to the historian as to how best to approach famine in societies of the past. The contradiction is only superficial. De Waal regards dependence on documents as suspect, preferring to base his account of famine in the Sudan on oral information culled from the people on the ground, and on his own observation. This approach recalls Finley's scepticism in the face of the primary sources. The crucial difference lies less in attitude to the sources than in the range of sources available to the historian of past societies, on the one hand, and the social scientist observing contemporary societies, on the other.

As to how broad one's definition of famine should be: de Waal's approach is comparable to that of the 'famine as process' theorists referred to earlier, who would apply the term famine to the whole course of a food emergency rather than merely to the last stage; he would, furthermore, call those food emergencies that stop short of a catastrophic last stage, famines, not merely those 'that kill'. Whether or not we as historians choose to follow his lead in this is perhaps unimportant. We are not 'disaster visitors' who, for better or for worse, will decide the fate of Sudanese or Ethiopians. De Waal's most useful contribution for our purposes is his subtle analysis of the evolving strategies of a population in the grip of a food-crisis. Along the way, he makes the kind of distinctions between one food-crisis and another that we are interested in making, without being able to draw on the perceptions of the suffering populations themselves.

For ultimately it is more worthwhile to ask whether one food-crisis was less serious or more serious than another, and what kind of structural differences are revealed by the comparison, than to decide whether to confer or withhold the title famine from either or both. In asking the former questions, we are exploiting the advantages of the food-crisis continuum: we are in effect engaged in placing a particular food-crisis on the continuum both in relation to other crises, and in relation to the two poles. Is it to be placed towards the famine end of the continuum, or alternatively towards the mild-shortage end? In the section that follows I ask by what criteria we can make such judgments. In the meantime I stand by my strong definition of famine, which in any case seems implied by the concept of a food-crisis spectrum, which opposes a 'famine end' and a 'shortage end'. My response to those who prefer a broad definition is to point to the advantages for a comparative analysis of making firm unambiguous distinctions. And every historian of famine must be a comparativist in some degree.

## CRITERIA OF FAMINE: MORTALITY INCREASE

My provisional definition of famine involves a sharp rise in the mortality rate and social, political and moral collapse.

Collapse of the social, political and moral order is the central, also the controversial, element of my definition. I come to it in the final section. Dramatic increase in mortality, by contrast, is a standard defining characteristic of famine, heading most people's lists. In the 'revised version' of my definition, I make it a leading symptom of social collapse, but only one of several symptoms. There are two main problems in applying this criterion. The first is how to decide what might count, in the context of food deprivation, as a dramatic rise in mortality, a mortality crisis; the second is, how to distinguish between mortality increase that is caused by food deprivation and mortality increase with other origins.

**How many must die?** The question has little practical relevance for historians of pre-statistical societies, for example England before the second half of the sixteenth century when burial records in parish registers become available. But how are mortality figures to be used when they are available?

There is of course the familiar problem of imprecise, discrepant and manipulated statistics, which I do no more than mention here.

One possible index is absolute numbers of deaths; so, Ireland, the Great Potato Famine: at least 1 million dead; Bangladesh, 1974–5: around 1.5 million; Bengal in 1943: around 2 million; the Soviet Union, that is, Ukraine and other European provinces, 1933: 4–5 million; China in 1959–61: officially 15 million, in reality probably many more.

Another index is an increase in the crude death rate (CDR), the numbers dying per thousand of the population. One can plot the increase in CDR above a moving average of deaths over a designated period of time.

The most useful indices will be those allowing us to take account of differences in population levels as between societies.

But again, where is the famine line to be drawn? Must mortality rise by 100%, as Andrew Appleby suggested in his study of famine in Tudor and Stuart England? The consequence of applying that estimate would be almost to banish famine from English and European history. The Europe-wide food crisis of 1816–19, called by John Post 'the last great subsistence crisis of the western world' would be revealed as of less than famine proportions. The crisis of 1846–7 in the western Highlands of Scotland, dubbed by T. M. Devine 'the great highland famine', would lose its title. In English history, the so-called Great Famine of 1315–16, in which some towns are thought to have suffered mortality increases in the range of 8–20%, might have to be renamed. In early modern England we find mortality rates of 21% above the norm in 1596–7, 26% in the following year and 18% in 1623–4. These are of course national figures, disguising regional differences, roughly between the north, which suffered greatly, and the south, which suffered less. In the Great Irish Famine the impact of the famine was again uneven, but the death rate really did double over the nation as a whole.

How to escape the *reductio ad absurdum*: no famine unless the death rate has doubled? We could tinker with the figures; we could say that mortality must rise not by 100%, perhaps, but by 50%, or 25%, for there to be a famine. Any new figure would be just as much open to the charge of arbitrariness as the old one. It would be better to stop thinking in terms of a numerical cut-off point, for behind that approach lies the unrealistic assumption that there is a sharp dichotomy between famine and lesser food crises.

Wrigley and Schofield, the historical demographers of early modern England, arrive at a workable compromise; they employ a one-star, two-star and three-star classification, corresponding to a mortality rise of, respectively, 10–20%, 20–30%, and over 30%, above a moving 25-year average.

Their concern is with ranking mortality crises in general (for the years between 1541 and 1871), not hunger-related crises in particular; and in fact the graver crises in their list were not famines at all. This leads me to the second problem to do with the mortality criterion of famine: how to distinguish between mortality crises induced by famine, those induced by epidemic disease, and mixed crises.

**Famine and disease** The two often strike together and are hard to disentangle in the evidence. In ancient Greek, the words themselves are easily confused: *limos* is severe hunger or starvation, *loimos* epidemic disease. In Greek literature, the two work in harness. Starvation and disease are the punishment of Zeus for the violent and evil city in Hesiod's poem *Works and Days*; in Herodotus' *History*, they together pursue the Cretans returning from Troy.

Of course, the ancient Greeks could tell apart food crises and epidemics, if for example people were stricken with disease in the absence of food crisis – as in fact happened in Athens in the second year of the great war of 431–403 BC between Athens and Sparta, described by Thucydides. Further, the copious medical writings that survive from antiquity provide the basis for a distinction between death from starvation, death from hunger-related diseases (especially dysentery and 'famine diarrhoea') and death from infectious diseases acting independently (measles, smallpox, plague, etc.), even if the distinctions were not yet fully conceptualised. For example, Galen, the doctor/philosopher from Pergamum in Asia Minor in the second century AD, knew that starvation was a rare form of death ('people do not die primarily because of want of food'), and found the main cause of death in famine conditions to be what we might call secondary infections. He has a colourful description of a peasantry succumbing to such diseases (the identities of which are not uniformly easy to determine) following the consumption of unwholesome substitute foods:

> The food crises (*limoi*) occurring in unbroken succession over a number of years among many of the peoples subject to the Romans have demonstrated clearly, to anyone not completely devoid of intelligence, the important part played in the genesis of diseases by the consumption of unhealthy foods.... The country people finished the pulses during the winter, and so had to fall back on unhealthy foods during the spring; they ate twigs and shoots of trees and bushes, and bulbs and roots of indigestible plants; they filled themselves with wild herbs and cooked fresh grass ... And so one could see some of them at the end of spring and virtually all of them at the beginning of summer catching numerous skin diseases, though these diseases did not have the same form in each case; some were erysipelatous, some inflamed, some with a lichen-like growth, some psoriatic, and some of leprous character ... But with several of them anthrax or cancerous tumours occurred along with fevers, and killed many people over a long period of time, with scarcely any surviving. Numerous fevers occurred without skin diseases; defecation was evil-smelling and painful, and there followed constipation or dysentery; the urine was pungent or indeed foul-smelling, as some had ulcerous bladders. Some broke out in a sweat, evil-smelling at that, or in decaying abscesses. Those to whom none of these things happened all died either from what was clearly inflammation of one of the intestines or because of the acuteness and malignity of the fevers.

Galen did not appreciate the finer points of the synergistic relationship between malnutrition and disease. He was, however, dimly aware that malnourishment can actually be a protection against certain epidemic diseases, such as malaria. This is now common knowledge.

Among modern historians, the complex interactions between nutrition and disease have been explored mainly by historians of early modern England. The Tudor and Stuart historian Andrew Appleby did pioneering work in disentangling the causal roles of epidemic disease and food deprivation (he called the latter famine), and in exploring the connection between the two. Recent research has taken considerably further the investigation of the relationships between nutrition and disease, and between, on the one hand, high grain prices and, on the other, food availability, famine, deaths from epidemic disease and mortality crises in general. This research is tending towards three main conclusions. I summarise the discussion of Walter and Schofield and, to avoid confusion retain their terminology of famine, famine-related mortality, and so on, rather than substituting my own. These conclusions are:

1. Mortality fluctuations in early modern England and France were overwhelmingly determined by the prevalence of epidemic diseases, not by famine.
2. Even in mixed crises, in which grain prices and the incidence of diseases appear to fluctuate independently, the role of famine is ambiguous.
3. The effect of famine is largely indirect. Famine causes dislocation; it stimulates migration from one country centre to another, or from the country to already crowded and unhygienic cities; and it is in this way that infection, the real killer, is spread.

This last observation applies more pertinently to shortage than to famine. Indeed, the discussion of Walter and Schofield as a whole contains an implicit contrast between, on the one hand, 'standard' harvest fluctuations, which provoked, *inter alia*, disease-spreading migration, and, on the other, exceptional shortage, which as in Galen's vignette of the peasants of Asia Minor induces life-threatening action, including the consumption of bad food or non-foods. This is my own shortage/famine distinction in embryonic form.

The net result of this exploration of the causes of mortality crises is that the role of famine is reduced and its historical presence in general diminished. This supports my earlier suggestion that famine is a comparatively rare event. More narrowly, it suggests that too much weight has been put on the requirement of a dramatic rise in mortality. Other criteria of famine, including other demographic variables, should be introduced to share the burden of defining famine. A full discussion would include the impact of food crisis on fertility. I confine my remarks here to migration.

**Migration** This is commonly seen as a safety-valve, reducing the death toll at home. But there are three qualifications to be made: first, migration does not always hold much hope for the migrant; second, it is not always possible; third, it may have adverse effects on the home community.

The first point needs no extended discussion: migration may not improve the prospects of a migrant, where for example he is destined to die in a refugee camp or migrant ship.

Opportunities for migration may be limited, especially in the case of the poorer members of an afflicted community, those most at risk in a food crisis. Or consider the ancient Egyptians. In a food crisis the best they could do was to move up or down the Nile into another administrative district. But that district might be expected to be in difficulties also, if the source of the problem was the absence of a Nile flood and therefore the lack of opportunity to plant a crop. The desert offered no escape to the fellahin. Moreover, in some historical periods at least, for example, the period of Roman domination, the route into the Mediterranean was closed to those without a special government-issued permit. Egyptians had nowhere to go – as a race. High levels of mortality in famine conditions are predictable, and stories of cannibalism become intelligible against this background.

Finally, mass migration can have negative consequences. Consider the emergency in the Scottish Highlands from 1846. The death toll was so modest that the historian of this famine felt obliged to pose the question whether it was a famine at all. But what happens if large-scale emigration is introduced as an index of famine? Outmigration from the Highlands was already under way before the first of the failures of the potato harvest in 1846, but it accelerated in the decade that followed. (Similarly, in Ireland, migration was considerable in the three decades preceding the potato famine and was then stepped up.) It was migration that enabled the Scottish Highland communities to avoid a mortality crisis of Irish proportions. But from the point of view of the home community, the results in Scotland and Ireland may not have been so dissimilar. Emigration reduced the population of the West Highland parishes by one quarter to one half in the 1840s and 1850s. In short, emigration might operate not only as a safety valve but also as a destroyer of communities. In this latter, destructive, role, migration deserves to be joined with mortality as a defining characteristic of famine. One can still agree that the Scottish famine was much less serious than the Irish famine, in which at least a million died and over 400 000 emigrated. But in making this judgment I am not relying solely on mortality statistics.

There are other criteria for identifying famine (and assessing the relative gravity of food crises) in addition to mortality and migration. But rather than fragment the definition I prefer to subsume all such criteria under what I regard as the central defining characteristic of famine – that it entails the social, political and moral collapse of the community.

## CRITERIA OF FAMINE: FAMINE AS BREAKDOWN OF A SOCIETY

Any reasonably stable community has evolved a system of political authority, social relationships and economic mechanisms for distributing available food throughout the population. It is human decision and action which shore up this system or alternatively undermine it, causing crisis, and in the worst cases, famine. Famine is a catastrophe that no society or polity can survive, at least in the short-term.

Symptoms of a disintegrating society include: political corruption, instability or anarchy, economic dislocation, descent into lawlessness and disorder, and the breakdown of the moral economy – all this resulting in severe food deprivation as indicated by the recourse to last-resort substitute foods, heavy loss of life, and large-scale emigration. In what follows I focus on the political dimension of famine and on the moral economy, and I play off these factors against two others that bulk large in most accounts of the aetiology of famine, namely, natural disasters, and, more briefly, prevailing economic and environmental conditions.

**Famine and politics** Famine has occurred periodically in times of political disruption. In particular, there has always been a close causal link between war, foreign or civil, and famine. One thinks of contemporary events in the horn of Africa or of the sufferings in the Second World War of Russians, Dutch, Greeks, and of Polish Jews in the Warsaw Ghetto. Historians of classical antiquity are aware that many famine narratives are set in times of war and culminate in the starvation of a population under siege – Athens in 403 BC at the end of the war with Sparta, Athens again under siege from the Romans under Sulla in 83 BC, Rome in the midst of civil war in 42–36 BC, and so on.

Political ideology has fathered famines. Stalin starved the peasantry of the Ukraine in 1932–3 by requisitioning their grain to feed the industrial workforce in the cities and to pay for imports of machinery and ferrous metals. Mao Zedong's Great Leap Forward killed more than 15 million Chinese in 1959–61 by inducing a sharp fall in supply, milking the countryside to the advantage of urban consumers, and smashing the bureaucracy. The Chinese bureaucracy had played a vital role in earlier times in preventing famine, notably in the eighteenth century, which has been called 'the golden age of famine relief' in China.

The relation between political stability and the welfare of the population was unusually intimate in Egypt. The Egyptian economy, until quite recently, depended on the ability of the central authority to monitor and control the flooding waters of the Nile through a complex network of irrigation channels, and to store the surplus of good years against the inevitable but unpredictable bad years, when sowing was restricted or prevented by a too low or too high flood. Famine in Egypt is often associated with rulers who were corrupt, inefficient or preoccupied with domestic or foreign enemies.

In the history of Europe, governments have created or aggravated food emergencies through a policy of minimal intervention, while requisitioning grain or permitting it to be exported. Peripheries of empire are vulnerable to such treatment. One could illustrate from the Roman empire, with reference to the peasants of Asia Minor whose sufferings were witnessed by the doctor (and social historian) Galen, or from a strange happening of AD 99 when grain was shipped from Rome to Egypt, that is to say, *sent back* from the parasitical imperial capital to its place of origin. Or one could illustrate the point from Finland, decimated by famine in 1696–7, while its Swedish overlords permitted traders to sell abroad the wheat surplus from Scania; or again, from Ireland between 1846–8 under the direct rule of Westminster. The British government, guided by its famine administrator Charles Trevelyan, high priest of the new orthodoxy of 'sound political economy', distributed food to the 'incorrigibly indolent' Irish belatedly and reluctantly, and only after the failure of its preferred policy of creating jobs. In addition, the export of grain out of famine-stricken areas was tolerated. Imperialist attitudes and economic dogmas combined to undermine the old ideology of the moral economy.

**The moral economy** The term was coined by E. P. Thompson to refer to a consistent, traditional assumption on the part of the mass of consumers that their rulers were morally obliged to protect them from starvation. Some governments advertised their commitment to the moral economy by issuing official regulations – examples are the Famine Codes of India of the 1880s, and the Books of Orders of

## ORDERS ISSUED IN TIME OF
## SHORTAGE, 27 DECEMBER 1586

*Orders devised by $y^e$ speciall*
*comma(n)ddment of $y^e$ Qu. $Ma^{ty}$ for $y^e$*
*releiff and ease of $y^e$ present derth of*
*gray(ne) $w^{th}$in $y^e$ realme.*

Ye shall buye noe corne to sell it agayne.

Ye shall neyther buy nor sell any mann(er) of corne but in the open market, vnlesse the same be to pore handiecraftesmen or dayelaborers . . .

That the Justices of the peace $w^{th}$in their seu(er)all divisions haue speciall regard that engrossers of corne be carefully seene vnto and severely punished accordinge to the lawe . . .

That they take order $w^{th}$ the comen bakers for the bakinge of Rye, barlie, pease and beanes for the vse of the pore . . .

That the justices be straightlie comaunded to see by all good meanes that the able people be set on worke, the howses of Correction provided and furnished and there ydle vagabonds *to be* punished.

*Figure 11* Orders issued in time of shortage, 27 December 1586.

Elizabethan England (Figure 11). In other societies, government involvement might be minimal. In the cities of the ancient Mediterranean world, the alleviation of food crises was characteristically undertaken by private benefactors, not by governments. In rural England of the early modern period, the wealthy were expected as a matter of course to provide grain to the poor in times of dearth. In addition, charitable institutions funded by state, church or private foundations play an increasingly active part in European society from the late middle ages, as we witnessed earlier in the case of fourteenth-century Florence.

To round off this sketch of the moral or social economy, it may be noted that in addition to the vertical links between rich individuals or institutions and the poor, there operated in traditional societies a horizontal support system between kin, neighbours, friends within the community and further afield. These links were fundamental; they are less often discussed because they are less well documented.

The same phenomena can be analysed, following Amartya Sen, in terms of entitlements, where entitlement refers to the basic ability of individuals to command the food resources they need for survival. Sen's theory is primarily an economic theory. What he called the exchange entitlement of an individual was measured primarily in terms of occupation and place in the network of economic relationships. Access by agricultural labourers to food hinges on their capacity to sell their labour power, which is much less in demand in a food emergency. Craftsmen and traders must buy food at high prices at a time when the demand for their own products and services has sharply declined. And so on.

In traditional societies, people do not depend entirely on the market for their food. As has been mentioned, a complex relief system operates at several levels. Translating into the language of entitlement, we can talk in terms of 'dependency entitlements' (vertical links) complementing 'reciprocal exchange entitlements' (horizontal links).

In a famine, this structure breaks down. The horizontal support system loses its utility: destitute peasants have nothing to exchange. Vertical links between large landowners and their tenants, labourers and other dependents prove fragile. A major theme of Scottish and Irish famine history is the mass eviction of peasants as a cost-saving exercise. As for public relief, governments of regions that are chronically vulnerable to food emergencies are commonly unable or unwilling to cope effectively with famine. There is little the poor can do in a famine to stimulate public authorities into relief action. Grain riots are a hallmark of shortages, not famines. It has been provocatively asked whether hunger rioters in early modern England were actually hungry. The point being made is that such riots were in part political, designed to remind the authorities of their obligations under the moral economy. In a famine, unrest is ineffectual and shortlived; it soon gives way to the desperate search for substitute foods, and then to apathy:

> Great destitution at Oranmore, county Galway: Whole families are living on chicken-weed, turnip-tops, and sea-weed; they did not ask for anything, no one spoke, a kind of insanity, a stupid despairing look, was all that was manifested.
>
> *Dublin Evening Mail*, September 1847

> They sit on wooden benches, crowded close together and all looking in the same direction, as if in the pit of a theatre. They do not talk at all; they do not stir; they look at nothing; they do not appear to be thinking. They neither expect, fear nor hope for anything from life.
>
> A poorhouse in Dublin through the eyes of De Tocqueville

## FAMINE: THE ROLES OF NATURE AND OF MAN

My account of famine places a heavy burden of responsibility for famine on people. For this reason, famine is, among catastrophes, unlike earthquakes or volcanic eruptions, which are natural disasters. However, Nature, and in addition, underlying economic structures, have traditionally been assigned leading roles in famine-genesis.

Now, humanity is heavily implicated in the deterioration of the natural environment. It is unnecessary to dwell on this obvious, and in the contemporary world, frightening, truth. In addition, the causal links between natural disaster, economic backwardness and famine are complex, and human action or inaction almost invariably has to be built into the sequence. This point does need some exegesis.

Rhys Carpenter argued that drought killed off the civilisation of Mycenae around 1200 BC. Barbara Bell offered a similar explanation for the disappearance of other ancient societies.

The onus is on the drought theorists to establish the existence of cycles of drought, and to show that these cycles had the enormous impact on history that is postulated. Instead, Bell attempted to prove the existence of the supposed droughts from the historical record. The attempt was not successful, and in any case it is a difficult and dangerous game to play.

Climatologists have suggested that the Mycenaean region might have suffered a run of unusually dry winters comparable to that of the winter of 1954–5, when the rain-bearing cyclonic depressions passed over Greece about 100 miles further north than usual. At most they have provided a model of what might have been. This is far from a demonstration that things happened in that way, and it is hard to see what pertinent evidence could be adduced. Phenomena such as droughts leave little trace in the geological record. Techniques such as tree-ring analysis might in time have something to offer; bad years show up as sequences of less than average growth. However, even if cycles of drought at the appropriate times could be pinpointed, it would still have to be shown that such conditions, through their impact on human society, changed the course of history.

The difficulties may be briefly illustrated from Turkish West-Central Anatolia between 1560–1620. An enterprising investigator has found suggestive sequences of years of restricted tree-growth in this region at this time. Putting this evidence together with scattered literary references, he presents a bleak picture of 'shortages' and 'famines'. However, the dimensions of the tree-rings cannot be shown to correlate closely with the position of consumers in Turkish Anatolia. For example, prohibitions on grain export, a sure sign of shortage, operated in 1565–7, but in those years above-average or near-average tree-growth is reported. This is an invitation to introduce non-climatic factors into the causal sequence. The period in question is acknowledged to have been one of 'peasant unrest, even revolt (the so-called Celali Rebellions), large-scale changes in land use, and unexpectedly large fluctuations in urban populations'.

It can be agreed that short-term climatic changes *may* fatally undermine humanity's capacity to manage the environment, especially in areas where the ecological balance is inherently precarious. The case is not yet proven for Mycenaean Greece or early modern Turkish Anatolia. But Cambridge historian John Iliffe, in his study of poverty in Africa, has drawn attention to the half century or so of natural disasters in pre-colonial and colonial Africa from the 1880s. The prolonged droughts and severe environmental degradation of Africa in the 1970s and 1980s, not to mention flooding in Bangladesh, are too familiar to need extended comment.

The question is, how exceptional such circumstances have been in human history. I suggest that Nature, whether in the form of climatic irregularities, environmental stress or crop disease (as in Ireland and Scotland in the 1840s) is more commonly the proximate cause of famine than the final determinant.

Nature may trigger off famine. But there is no straightforward causal chain leading from climatic irregularity through harvest failure and food shortage to famine.

Harvest failure, however caused, is not a necessary cause of famine. Sen noted that the Great Bengal Famine of 1943 followed a harvest only 5% down on a five-year average for the region. Famine can occur, and usually does occur, when there is no absolute food shortage. Nor is harvest failure a sufficient condition of famine. There was no famine in Bengal in 1941, although the harvest was 13% lower than that of two years later, the year of the Great Famine. Even in ancient Egypt, the combination of efficient storage and distribution systems could counter one bad year or several; this appears to be the message of the Joseph story in Genesis. The example should not be pressed. Pharaoh had a Joseph to interpret his dreams. Would he have stored so much grain otherwise, and over so long a period (though a drought of precisely seven years is not an authentic detail)? And would it have been reasonable to expect him to do so?

I turn finally and very briefly to the role of underlying economic structures. In famine, the link is broken between a community and the resources that ordinarily sustain and reproduce it. This has been a fragile link, easily jeopardised, in numerous societies, for example, in the traditional underdeveloped societies of the pre-industrial West, or in modern developing countries. Such societies are characterised by low productivity in agriculture, primitive transport facilities, poverty and low entitlement among the mass of consumers. I prefer to put it in these terms, rather than say, following Malthus, that famine follows from the conjunction of too little food and too many mouths to feed. However, people have not been passive in the face of recurring, though not precisely predictable, environmental disasters and economic setbacks. They have evolved flexible long-term strategies for the survival of their societies. History has shown how easily communities slip into shortage, hunger and destitution, and how determinedly they resist the plunge into famine.

## CONCLUSION

Thus, if we are to ask why famine occurs, or occurs more commonly in one society rather than another, we can begin with the environmental background; we can pass to the economic system and consider the level of agricultural productivity, the transport system, and the degree of market integration. But it is necessary to go beyond these factors to investigate the social or moral economy, the effectiveness of private relief systems in insulating the more vulnerable sections of the community against starvation, and the supplementary role of institutions and governments. This is an essential part of the explanation of how, for example, England slipped the shadow of famine by the end of the eighteenth century, leaving Scotland and Ireland behind; or how China in the eighteenth century was relatively famine-free.

If today the poorer countries of the world could in principle be protected from famine, this is not only because there is enough food to go around, but also because a supra-national relief system, a world moral economy, has come into existence, capable of staving off famine – if only local governments will co-operate, and if only relief organisations will diagnose correctly the needs of the communities at risk.

This is a notable advance. There remain, however, those food emergencies that are terrible enough from the point of view of the victims, but fail to engage the conscience or even attract the notice of the nations of the world, because they are not 'famines that kill'. Finally, one can ask whether the main opponent has been identified. Is it famine, or not rather malnutrition? It may be in our power to abolish episodic famine. Endemic hunger, the constant or continuing catastrophe, remains.

## FURTHER READING

Arnold, D., *Famine: Social Crisis and Historical Change*, Oxford: Basil Blackwell, 1988.

Devine, T. M., *The Great Highland Famine: Hunger, Emigration and the Scottish Highlands*, Edinburgh: J. Donald, 1988.

De Waal, A., *Famine that Kills: Darfur, Sudan 1984–1985*, Oxford: Clarendon Press, 1989.

Harrison, G. A., ed. *Famine*, Oxford: Oxford University Press, 1988.

Newman, L., ed. *Hunger in History: Food Shortage, Poverty and Deprivation*, Oxford: Basil Blackwell, 1990.

O'Grada, C., *Ireland Before and After the Famine: Explorations in Economic History, 1800–1925*, Manchester: Manchester University Press, 1988.

Sen, A., *Poverty and Famines: An Essay on Entitlement and Deprivation*, Oxford: Clarendon Press, 1981.

Walter, J. H. and Schofield, R., eds. *Famine, Disease and the Social Order in Early Modern Society*, Cambridge: Cambridge University Press, 1989.

# The case of consumption

ROY PORTER

Panics shape destinies. Throughout history, the great fears have been great precisely because they fuse direct, tangible objects of terror with fantasy tissues of phobia. Amongst the most real of the catastrophes to hand is disease. We have long lived in the shadow of cancer. Now the threat of AIDS looms over us, reminding us that, even today, lethal diseases still constitute terrifying hazards to our very existence. When, as with cancer and AIDS, there is no cure to hand; and above all, when, as with cancer and with AIDS till recently, the very aetiology of the disease is obscure – under such circumstances, as Susan Sontag emphasises, deadly diseases spawn sinister connotations. They become associated not just with suffering and death, but with evil, divine retribution, plots and conspiracies, enemies within and the sickness of the soul. Medical ignorance begets moral panic, and we start speaking of cancers of society.

Bubonic plague, leprosy, syphilis, cholera – all these, because they were new, or sudden, or epidemic, inexplicable, or incurable, or specially physically disfiguring, were in their turn read as the stigmata of vice and sin, whether individual or collective. Over a span of nearly four hundred years, however, from the mid-sixteenth century to the mid-twentieth, it was another disease that perhaps proved the greatest catastrophe, the greatest catalyst of political perturbation in the Western World. Its symptoms? As described by the eighteenth century English clinician, William Heberden, they were 'shortness of breath, hoarseness, loss of appetite, wasting of the flesh and strength, pains in the breast, profuse sweats during sleep, spitting of blood and matter, shiverings succeeded by hot fits, with flushings of the face, and burning of the hands and feet, and a pulse constantly above ninety, a swelling of the legs, and an obstruction of the menstrua in women'. Its outcome? A protracted wasting unto death. Its name? Classically 'phthisis', anglicised to the 'tisick', or consumption, or, technically, in medical jargon, tuberculosis: above all, pulmonary tuberculosis.

TB – the 'white plague' – became for a couple of centuries the greatest single adult killer on both sides of the Atlantic. What made it so challenging was that it was so obviously intimately linked with, maybe even integral to, the march of progress. Apparently insignificant in the medieval countryside, consumption had clearly worsened from the sixteenth century with the onset of rapid urbanisation, decimating the commercial cities of northwest Europe, and then becoming the classic disease of the industrial world. Terrifyingly, tragically, it felled not those full of years, nor even babies too young to be mourned, but teenagers, breadwinners, and young mothers, those in the prime of life, the hopes of the future. Moreover, it struck all ranks. Unlike typhus, it was not a 'filth disease', afflicting only the huddled masses – and hence a disease polite society could safely ignore. On the contrary, the affluent and accomplished, the bright and the beautiful, were equally stricken – tuberculosis, of course, claimed more than its fair share of artists and writers, from Keats to Kafka. It typically proved fatal; and, if, in its earlier stages, not the most painful of afflictions, its lingering nature meant that the lives of kin and neighbours were long attended by a living death, a *memento mori*.

Unlike plague, which receded from Europe from the end of the seventeenth century; smallpox, decisively on the retreat after the early nineteenth century introduction of vaccination; cholera, which hit Europe in only three or four brief, if devastating, waves; and waterborne infections such as typhoid, which public health and civic

*Figure 1* Calmette developed the BCG vaccine, the introduction of which into Britain was delayed essentially for chauvinistic reasons.

improvements had on the run from mid-Victorian times, tuberculosis was still exceedingly virulent in the first decades of the present century, carrying off some 40 000 victims a year in Britain. Hindsight shows there were no decisive *medical* breakthroughs before the advent of the BCG vaccine (named after Calmette and Guérin, developed in France, and, perhaps for chauvinistic reasons, introduced only tardily into Britain), and the development of streptomycin, part of the antibiotic revolution after the Second World War. By then, however, improved housing and nutrition, and tuberculin-tested milk, were already getting the upper hand. Even so, of the five millions worldwide who died each year of TB around 1950, a substantial proportion lived in the West. Thirty years later, the global number of deaths remained about the same, but almost all came from the Third World. TB has a fair claim to be called the longest-running, the most serious, of all our disease catastrophes.

## BODIES NATURAL AND BODIES POLITIC

It is not my aim here to trace the rise and fall, or conquest, of tuberculosis: that has often been done, classically by René and Jean Dubos's *The White Plague*. Rather, I intend to explore this mysterious, and long deadly, affliction as a carrier of cultural connotations. What was the *meaning* of this catastrophe, individually and socially? What was this pallor, this exhaustion, this wasting, trying to say about the health of society itself?

In posing this question, I take two cultural truths for granted. First, as already mentioned, deadly epidemics beget moral mythologies. Second, the human body has always been the richest of metaphorical resources. From Plato's *Republic*, through the cosmology of macrocosm and microcosm so dear to the Renaissance, down to *fin de siècle* degenerationism, the fate of the human body, the body natural, has been mapped upon the body politic – indeed upon Creation at large. The health of the individual reflects the well-being of society in general; the distempers of the body or mind mirror and match social disorder. Notions of excess and deficiency, conflict and chaos, circulation and stagnation, activity and repose, struggle and equilibrium, vitality and waste, permeate discourse on the order of things, individual and social, and inform social pathology and psychopathology. No disease is experienced as catastrophic, without, by implication, threatening, or registering, social and personal catastrophe too. It is in the light of these natural beliefs that I shall examine how the disease of consumption became a symptom of, a sermon for, a society increasingly devoted to consumption, a consumer society – and it needs no emphasising that the very term 'consumption', as well as labelling a disease, is central to political economy, and – in forms such as 'consuming passion' – to moral philosophy no less. What did the disease 'consumption' mean in the emergent consumer society? I shall proceed schematically, freezing time's flow around the close of three centuries: at 1700 (which I shall call 'early modern'), at 1800 (which I shall call 'late Enlightenment'); and at 1900 (which I shall call 'modern'), though my divisions will be more thematic than temporal. I will necessarily be highly selective in choosing representative texts.

## EARLY MODERN

How did people construe the flourishing state, and the healthy body, in the early modern period? As historians of economic thought stressed, wealth theory had long broken with crude 'bullionism' (the miser's dream that treasure lay in hoarding gold and silver), in favour of subtler theories privileging circulation: true wealth sprang from economic motion, stimulating labour, industry and exchange, a view perhaps acknowledging William Harvey's discovery of the circulation of the blood. Yet due measure had to be achieved. For one thing, prestigious teachings warned that excessive wealth was the cancer of the commonwealth. Churches anathematised the love of lucre and unbridled appetite, while civic humanism prophesied that private enrichment sapped public liberty and virtue. 'Luxury', warned John Dennis in 1711, is the 'spreading Contagion of which is the greatest Corrupter of Publick Manners and the greatest Extinguisher of Publick Spirit'. Mercantilism itself feared wealth would easily turn to waste. Herein lay a conundrum. Buying and selling were necessary for life-giving commerce: yet what was spending but the dissipation of accumulated resources? Conspicuous consumption was conspicuous waste.

It was especially problematic because, in a culture habitually interleaving the body politic with the body human, the road to health for the human body was itself difficult to find. Traditional theories of life likened vitality to a candle. In this decaying, sublunary sphere, the natural tendency of the flame to sink and die could be postponed by renewed external stimuli. Eating and drinking fuelled the vital fires.

Where hunger, dearth and even famine stalked the land, people needed little persuading that hearty eating and drinking protected against disease, debility, and death. Rhythms of work and rituals of sociability centred on celebrations, cheer, conviviality and community, by affording times and sites devoted to bingeing: the flowing

*Figure 2* The two greatest men. It was common in this period to parade obese specimens as if they were livestock or fatstock.

bowl of harvest home, *mardis gras*, wassailing, and wakes. Cuisine and cellar loomed large in the pursuit of happiness.

But they also figured in the quest for health; for traditional wisdom regarded high living as a form of preventive medicine. The Englishman's cultic roast beef was not mere chauvinism, gluttony, or fantasy, but positively therapeutic. 'Man is an eating animal, a drinking animal, and a sleeping animal', defined the extremely corporeal Dr Erasmus Darwin. Such was his own sesquipedality of belly that the great inventor ingeniously had a semi-circle sawn out of his dining-table, so that he could station himself within eating distance.

The stomach – that 'grand Monarque of the Constitution', according to Edward Jenner – needed to be active, to digest the copious quantities of aliment required to concoct the blood, spirits and humours enlivening the limbs. Hence the ideal victuals were savoury and strong, and the red meat and wine diet of the rich evidently more invigorating than the insipid gruel and water of the poor. 'My Stomach brave today', purred Parson Woodforde in 1795, 'relished my dinner'. Rarely a day went without logging his menus. Indeed, Woodforde's ultimate diary entry before his death culminates with a last supper:

> Very weak this Morning, scarce able to put on my Cloaths and with great difficulty, get down Stairs with help ... Dinner to day, Rost Beef etc.

John Locke was told by one of his friends that his wife, 'in order to her health ... is entered into a course of gluttony, for she is never well but when shee is eating'.

This model of the healthy body as a vital economy, requiring energetic stimulus, was widely accepted by the medical profession itself. Dr Thomas Trotter endorsed the advice of the famous Venetian longevist, Luigi Cornaro, who prescribed at forty, two cordial glasses of wine a day, four at fifty, and six at sixty, while Dr Peter Shaw wrote a book in 1724 to prove – to quote his own title – *Wine Preferable to Water*, indeed *A Grand Preserver of Health*. Alcohol was medicinal. For long, Parson Woodforde swigged a glass of port as 'a strengthening Cordial twice a day'. Prophylactic eating and drinking in turn required energetic waste disposal. Hence popular physiology attended to evacuations no less than to appetites. Purging was the panacea, but sweats and phlebotomy were important auxiliaries, emetics too.

Medicine thus conceived the pulsating body as a through-put economy whose efficient functioning depended upon generous input and unimpeded outflow. But how was this need for positive stimulus to be squared with equally venerable doctrines – both medical and moral – of temperance, moderation, and the golden mean? Might not energising the system precipitate pathological excess?

Such fears were often voiced, in the light of what an early Georgian pamphlet denounced as 'the present luxurious and fantastical manner of Eating'. Notorious for indigestible favourites such as pudding, the English were, as the saying went, digging their graves with

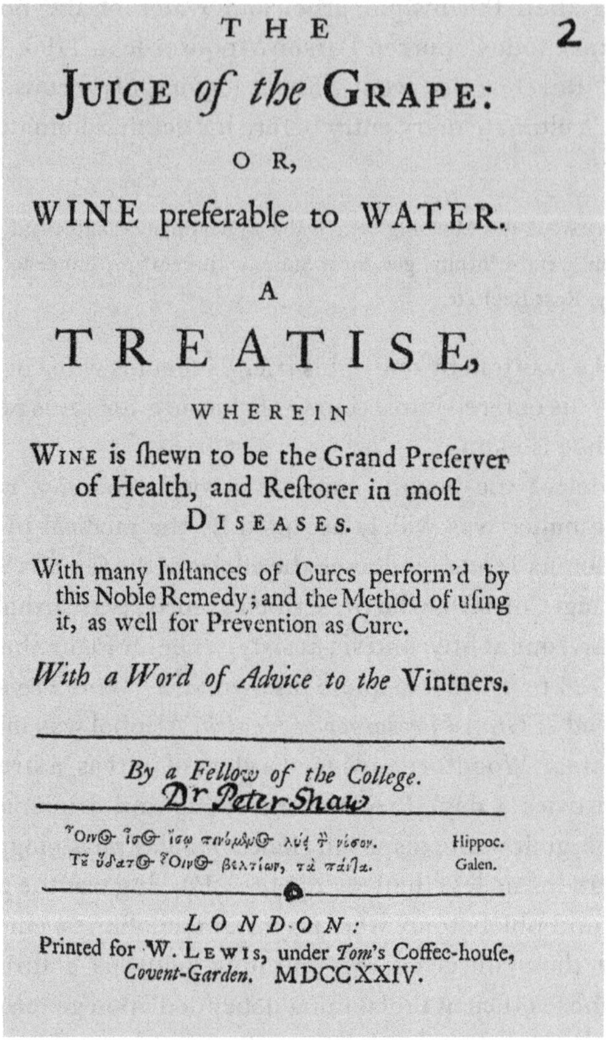

THE 2

JUICE *of the* GRAPE:

OR,

WINE preferable to WATER.

A

TREATISE,

WHEREIN

WINE is ſhewn to be the Grand Preſerver of Health, and Reſtorer in moſt DISEASES.

With many Inſtances of Cures perform'd by this Noble Remedy; and the Method of uſing it, as well for Prevention as Cure.

*With a Word of Advice to the* Vintners.

*By a Fellow of the College.*
Dr Peter Shaw

Οἶνος ἴσος ἴσῳ πινόμενος λύει τὴν νόσον. Hippoc.
Τοῦ ὕδατος ὁ Οἶνος βελτίων, τὰ πάντα. Galen.

*LONDON:*
Printed for W. LEWIS, under *Tom's* Coffee-houſe, *Covent-Garden.* MDCCXXIV.

*Figure 3* Juice of the grape. The health giving properties of alcohol were still widely touted in the eighteenth century.

their teeth. And if serious eating was parlous, gross drinking proved still more constitutionally lethal. Erasmus Darwin dubbed alcohol 'the greatest curse of the christian world'; and no wonder, for oceans were swallowed, and not just during the gin craze. Intoxication, judged Samuel Richardson, was 'the most destructive of all vices:

*asthmas, vertigoes, palsies, apoplexies, gouts, colics, fevers, drop-
sies, consumptions, stone,* and *hypochrondriac diseases,* are natur-
ally introduced'. Thus early modern medical gastronomics held that
appetite was healthy; but excess brought what Byron dubbed 'the
horrors of digestion'. Above all, protracted bingeing led to exhaustion, dyspnoea, dropsy, and even the 'tisick' or tuberculosis. Oliver
Heywood, the Nonconformist minister who delighted in detailing
delinquents' demises, thus recorded how a rival preacher, 'a bab-
bling, wretched creature', had first become a 'great drinker as its
sd[sic]', and had 'at last fallen into a consumption'.

And this is the point. Just as, in the kingdom, wealth readily turned
to waste, so in the individual, excessive consumption could, by a
paradoxical twist, produce not strength but dissolution. Late-seven-
teenth-century medicine was alarmed by the apparently rapid spread
of cachexies or wasting diseases – scurvy, cancers, scrofula, *tabes
dorsalis,* venereal infections, ascites, catarrhs, dropsy, asthmas and a
galaxy of hysterick fevers and hypochondriack melancholias. Gideon
Harvey explicitly designated consumptions as the '*morbus anglicus*' –
they were 'both an English *Endemick* and *Epidemick*' – which by
'devouring of parts' culminated in the 'corruption of the essentials'.
Benjamin Martin shortly pronounced: 'There is no Country in the
World more Productive of Consumptions than this our Island'.

All such scurvies and cachexies were diseases of wasting, exhibit-
ing symptoms including general malaise, weight loss, flaccid flesh,
poor skin tone, the failure of ulcers to heal, and a general 'rottenness'
of health. This congeries of chronic constitutional conditions was col-
lectively known as 'the consumptions', and included tuberculosis; it
was blamed upon excess. Feasting, drinking and 'sporting in the
Garden of Venus' were the prime causes of 'wasting diseases', resulting in what Thomas Willis called the 'withering away of the whole
body', down to a mere anatomy, and premature enfeeblement. Post-
Restoration physicians, such as Christopher Bennet, Gideon Harvey,
Thomas Willis, and Benjamin Martin, disputed about the pathological details but all were agreed in laying blame at the door of excess.
What was to be done? Traditional medicine commended the

ubiquitous therapeutics of 'evacuation' – diaphoretics and bleeding, 'repeated according to the strength of the patient and the present effervescence of the blood'.

In short, medical opinion argued that wasting conditions followed from growing opportunities for self-indulgence and high living. Wealth, ease, and urbanism were above all hazardous to 'the rich, who are not under the necessity of labouring for their bread'.

I have argued that the dialectics of wealth and waste perplexed early economists. I have gone on to suggest that they disturbed the doctors no less, for the enthusiastic consumer could himself end up being consumed, the hearty eater eaten. These apparent paradoxes were addressed by the early-eighteenth-century physician, George Cheyne, whose writings on chronic disorders proved particularly pivotal, encapsulating past wisdom, yet formulating new philosophies for the future. Cheyne wrote from personal experience. Cheyne was a Scot, trained at Edinburgh, who took the high road south early in his career, haunting London coffee houses and taverns, to drink himself into practice. He rapidly established a name as a witty man-about-town with 'Bottle Companions, the younger Gentry, and Free-Livers',

> nothing being necessary for that Purpose, but to be able to *Eat* lustily, and swallow down much *Liquor*; and being naturally of a large *Size*, a cheerful Temper, and Tolerable lively *Imagination*, ... I soon became caressed by them, and grew daily in *Bulk* and in Friendship with these gay Gentlemen.

High living eroded his health, however, and he grew 'excessively fat, short-breath'd, Lethargic and Listless'. Fearing for his life, he quit town, imposed an austere diet, and saw his corporation melt away 'like a Snow-ball in Summer'. Over the years, Cheyne's weight went up and down like a yo-yo. His worst crisis came around 1720, when, experiencing 'a Craving and insufferable Longing for more Solid and Toothsome Food, and for higher and stronger Liquors', he blew up to thirty-two stone, eventually needing a servant to walk behind him carrying a stool, on which to rest every few paces, until he 'went about like a Malefactor condemn'd', his gluttony producing 'Giddiness, Lowness, Anxiety, Terror', 'perpetual Sickness, *Reaching, Lowness, Watchfulness, Eructation*', and a nervous hypochondria which 'made Life a Burden to myself, and a Pain to my friends'.

Cheyne posed the key question: Did the wealth of nations secure the health of nations? Far from it. As England rose from rags to riches, her people sank in health, suffering uniquely from that clutch of chronic and constitutional conditions – 'nervous disorders' – which he dubbed 'The English Malady'.

Why so? Cheyne's explanation drew upon familiar primitivist tropes: 'when Mankind was simple, plain, honest and frugal, there were few or no diseases. Temperance, Exercise, Hunting, Labour, and Industry kept the Juices Sweet and the Solids brac'd'. All had changed:

> Since our Wealth has increas'd, and our Navigation has been extended, we have ransack'd all the Parts of the *Globe* to bring together its whole Stock of Materials for *Riot, Luxury*, and to provoke *Excess*.

As the nation grew 'luxurious, rich and wanton', so diseases mushroomed. The fatcats of town pursued opulent life-styles that only wealth would buy: soft beds, late rising, later nights, artificial lighting and heating, tight-lacing, and above all elaborate cuisines culled from all corners of the globe, involving dishes typically rich, salted, sauced, pickled, smoked, and highly-seasoned, all washed down with distilled liquors and ardent spirits. High living in high society carried high health risks.

Cheyne formulated the notion of the 'English Malady' as a constitutionally crippling yet socially eligible disorder – a badge of gentility. Three aspects of his advice need mention.

First, Cheyne set himself up as the apostle, if not of dietary abstinence, at least of austerity. He frequently put himself on strict regimes, sometimes totally abstaining from flesh and alcohol, and living on vegetables and water; he also commended comparable 'seed and milk' diets to other sufferers, especially those afflicted with '*a settled* Hectick *(from Ulcers), an* Elephantiasis *and* Leprosy, *a humorous*

Asthma, *a chronical* Diabetes, *an incurable* Scrophula *and a deep* Scurvy', to say nothing of *'higher and inconquerable hysterick and hypochondriack Disorders'*. Many latter-day vegetarians and 'low cal' zealots – not least Shelley – looked back to Cheyne's dietaries for inspiration.

Second, Cheyne challenged the credit of the 'high diet' by mounting a critique of corporeality, and replacing the carnal with a cult of what we might call 'the lightness of being'. Indeed, he formulated his own spiritualist Christian piety, drawing upon Platonic philosophy and Behmenist mystical immaterialism.

In particular – my third point – Cheyne advanced a life-style designed to refine the grossness of affluence into something more elevated and aetherial. He invented a new sociology (an aesthetics for elite living), a psychology (heightened sensibility, indeed, *taste*) and predicated them upon a new physiology, discarding classical humoralism for the iatro-mechanist idiom of the nerves, diverting attention from the traditional through-put economy of the fluids, and identifying the nervous system as the key to 'the Human *Machin*'. Excess clogged the nerves, rendering them sluggish and 'glewy', and causing ulcerations, inflammations, and other obstructions. Relaxed nerves would finally produce diarrhoeas, phlegm, spitting, rheums, dropsy, diabetes, scrofula, and so forth.

Fine physical health, it followed, depended upon keeping the nerves springy, clean, and tonic. Fine spirited people, living, as we say today, on their nerves, were particularly prone to nervous debility. Gross over-consumption had to be abandoned, and the tastes of the rich refined. To avoid the disease of consumption, it was necessary to avoid over-consuming.

## LATE ENLIGHTENMENT

I now move forward to my second slice of time, around 1800. It has been argued that there was a 'consumer revolution' in eighteenth-century England. At the same time, tuberculosis was also climbing to appalling heights. Around 1700, one in ten deaths registered in the

London Bills of Mortality had been attributed to pulmonary consumption. By its close, the figure was one in four. Tuberculosis had become the single largest killer of adults, in Erasmus Darwin's phrase, a 'giant-malady . . . which . . . destroys whole families, and, like war, cuts off the young in their prime, sparing old age and infirmity'. Did late Enlightenment medical opinion see a connexion between the features of the new consumer society and the dreadful spread of tuberculosis? It did, but the relation perceived was complex. Doctors did not see new consumerism simply as yet more 'excess', nor did they follow their predecessors around 1700 in blaming wasting diseases on crass personal over-indulgence. Rather, they thought that calls such as Cheyne's for refinement had ironically proved too successful, bringing about what Thomas Beddoes called changes 'in almost every circumstance of the manner of living' which had proved far more pathological. I wish to explore such developments, by turning to the critique mounted by Thomas Beddoes, a physician famous for his friendships with Humphry Davy and the Romantics, but (most germane here) distinguished for his investigations into tuberculosis made at the Pneumatic Institution he founded in Bristol in 1799.

Why was consumption spreading so catastrophically? Beddoes pinpointed factors we can see as central to McKendrick's 'consumer revolution'. One cohort of its victims was among the labouring population, in particular operatives in sedentary, indoor trades such as (a splendidly Smithian example) 'needle-grinders' – where workshop atmospheres were polluted with fibres, dust, and particles that irritated the lungs. These people succumbed not through inherited constitutional weakness, but because of the 'nature of the occupation'. Such workers were thus 'forced into the disease', through their own 'self-neglect' and the 'unconcern' of their masters. Beddoes singled out further changes imperilling the work-force. Men who abandoned physical hard labour, allured by good money in 'the almost feminine occupations of the cloathing manufacture', became 'frequently consumptive'. Worse still, all kinds of working people had been seduced into wearing the new fashionable textiles – 'light cotton dresses, instead of the warmer plaid which was formerly worn', which must bear the blame for 'no small share of the equally common prevalence of colds, fevers, rheumatisms, asthmas, consumptions'. Thus the producers of the 'consumer revolution' often paid the tribute of their health and even their lives.

But the consuming classes were even more consumed by consumption. Here Beddoes blamed the affluent for pursuing pernicious lifestyles, which sacrificed health to the household deities of fashion. Fetishising belongings and appearances, they neglected well-being. Crazed by fashion, for one thing, the rich clad themselves in the 'light dress' which was the rage in the Revolutionary '90s. This required fierce indoor heating, which in turn weakened people against the cold, creating chills and coughs and ruining the constitution.

Frivolous fashions were, however, but the tip of the iceberg. For the modish 'method of education' fostered in polite circles was almost custom-built to turn children into weaklings. Captivated by the new sensibility, parents forced their infants into study, music and fine accomplishments. Even girls were now packed off to be 'poor prisoners' in draughty boarding schools. Adolescents, 'weak, with excess of sensibility', were then allowed to loaf around on sofas, reading improving literature and 'melting love stories, related in novels'. Occupations designed to 'exercise the sensibility' proved 'highly enervating'. Not surprisingly, thanks to this 'fatal indolence', 'the springs of their constitution have lost their force from disuse'. Beddoes spied a further insidious danger in sedentariness: the solitary vice. Masturbation became the target of Georgian medical writers who exposed it not just as a sin, vice, or character weakness, but as ruinous to health, because the 'waste' or 'spending' of semen supposedly induced wasting conditions.

Worse still, by its association with modish sensibility, tuberculosis was becoming positively fashionable. 'Writers of romance (whether from ignorance or because it suits the tone of their narrative) exhibit the slow decline of the consumptive, as a state on which the fancy may agreably repose, and in which not much more misery is felt, than is expressed by a blossom, nipped by untimely frosts.' The preposterous idea had grown up, Beddoes alleged, that 'consumption

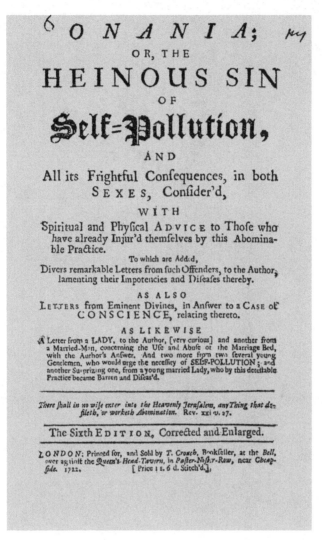

*Figure 4* Onania. Eighteenth century opinion believed that masturbation was a major cause of wasting disease.

must be a flattering complaint', because decline from the 'valetudinary state' was so gradual, initially painless, and non-disfiguring. Consumption conferred an intriguing, enticing languor, and had become associated with superior imagination, talents, and discrimination.

In the process whereby a girl was 'manufactured into a lady', parents positively encouraged delicacy. Because it was in to be thin,

*Figure 5* Morton. Morton wrote around the very peak of the
incidence of tuberculosis and its Romantic myth.

such creatures were allowed to become finical eaters. Hoodwinked by
pseudo-medical faddery – maybe they even read Cheyne! – many
parents encouraged vegetarianism, convinced it purified the blood
and the nerves: 'There are ... among the higher classes, some who
keep their children to the fifth, or even the seventh year, upon a strict

vegetable and milk diet', revealed the appalled Beddoes, 'believing that they thus render the constitution a signal service', seduced by the 'false hope of rendering the blood of their children pure, and their humours mild'.

An end to that! Beddoes urged a return to the heartier, pre-Cheynian days to counter 'the mistakes of those prudent parents who imagined that by confining their children to a vegetable diet, they were purifying their blood, while, in reality, they were starving them into scrophula'. 'It seems probable, that the general diet of former centuries was more invigorating', for the 'opulent of both sexes, appear to have participated rather more largely of animal food', often breakfasting 'upon a fine beef steak broiled' – so good for 'resisting cold'.

Thus precious life-styles left the elite specially vulnerable: 'it is upon the lilies of the land, that neither toil nor spin, that the blight of consumption principally falls'. 'Fixed . . . immoveably upon well-cushioned chairs and sofas, in hot, close apartments', they were too weak to 'receive and digest a proper quantity of aliment'. Far healthier were labourers who pursued heavy physical exercise, drank with gusto, and above all, were lusty carnivores.

As is obvious, Beddoes's arguments spell a remarkable turnabout from earlier medical discourse. Late Stuart doctors blamed consumption on excess; Beddoes indicted deficit. It is tempting to speculate that Cheynian arguments against 'high living' had helped induce the transformation Beddoes denounced.

Beddoes thus anatomised this 'giant-malady' as the product of an economy and a life-style, as a disease of civilisation. Proceeding 'from domestic mismanagement, and not from the inalterable dispositions of nature', it was preventable. Obsession with objects and indifference to health multiplied the 'tribute of lives we render to consumption'; for the desire 'of dazzling strangers by the splendour of an equipage or by the lights of the understanding' was 'so prevalent'. Where did remedy lie? 'We must all learn not to bear to have every thing about us – cloaths, tables, chairs, pictures, statues – all exquisite in their kind – except our progeny'.

So what was to be done? Beddoes condemned what he saw as a further evil sign of the times, a tendency amongst the rich to be blasé about tuberculosis because money could buy abundant medical services, and sufferers could be packed off to the spa, the seaside, or the South. Contemporary medical opinion recommended warmth to counter consumption, and Beddoes certainly had no quarrel with that, but he believed that the sending of patients to the South Coast, or to Portugal, Madeira or the Riviera was nothing better than getting the dying out of sight and out of mind – most convalescent resorts were no better than charnel-houses run by medical racketeers. Beddoes advised a superior way of providing comforting warmth to his patients. This was the 'cow-house method'. Beddoes urged consumptives to live in barns, with cattle for companions, till cured. 'Stabling with cows' is 'not unpromising', Beddoes insisted.

The philosophy of the cow-house was two-fold. The beasts themselves would yield regular, steady warmth, twenty-four hours a day. Moreover, the volatile alkali simultaneously exuded would help purify the lungs. Thus, waste products would prove not wasteful at all, but triumphantly integral to the economy of health, in a manner that would surely have given joy not just to Dr Pangloss but to that great Victorian sanitarian campaigner for recycling excrement, Edwin Chadwick.

Beddoes devoted much of his *Observations on the Consumptive* to evaluating this treatment which he had 'long been in the habit of recommending', even though, he confided, 'not unfrequently did I forfeit the good opinion of my patient'. It had, he claimed, enjoyed signal success, for instance in the case of Mrs Finch, none other than Joseph Priestley's daughter, who, in a letter headed 'Cow-House, *Oct 8*', explains that 'she has found a cow-house a much more confortable abode than she had formed an idea of', though the stench was 'nauseous' and 'successive generations of flies were a considerable nuisance': 'I am', she concluded, 'more than ever a friend to the cows'.

## MODERN TIMES

Let us turn, finally, to our third period. Around 1700, consumption was treated as the reward for excess. By 1800, I have suggested, it was punishment for refinement, or, in other words, for excessive consumption of an effeminising culture. How was it viewed by 1900? Here I shall be very brief, because this part of the story has recently been well told in Linda Bryder's *Below the Magic Mountain: A Social History of Tuberculosis in Twentieth-Century Britain.*

TB peaked in Britain around 1875, when some 50 000 people per year were dying directly from it – around one in eight deaths – and more were infected. In the succeeding half century, it still haunted the public health and political consciousness, as the disease of the working classes.

In 1882, Robert Koch isolated the tubercle bacillus responsible for consumption. Thereafter, it became in principle easy to detect the disease's presence. But this bacteriological advance did not speedily lead to a cure – despite Koch's early enthusiasm for tuberculin treatment. Nor did it dispel the age-old urge to moralise the condition. This time concern was focused upon the poor.

It was, of course, well known that industrial workers in certain trades were especially susceptible: Beddoes himself had pointed this out. The late Victorian medical profession, however, increasingly denied that such external, workplace, conditions should be held directly responsible. They were at most but triggers, highlighting the condition in those who independently had a hereditary disposition to the disease. This notion of a hereditary diathesis chimed, of course, with the much-touted theory that large segments of the lower classes were, and were bound to be, sick, precisely because they were, viewed within a grander social Darwinistic vision, defectives, deteriorated stock, even degenerates. Why did tuberculosis afflict numerous members of the same family? Not (it was argued) because it was contagious, spread through droplets, but because it ran in families; it was a genetic taint.

One leading strand of late Victorian and Edwardian medical

*Figure 6* Robert Koch. Koch was the discoverer of the tuberculosis *bacillus*.

opinion, held especially by those of a eugenist turn of mind, were convinced that it was not only such physical defects to which the residuum was hereditarily disposed. Widely accepted ideas of degeneration proposed that moral and personal weakness were also 'in the blood' as it were. And this provided a further explanation for the prevalence of tuberculosis. For why did such a high percentage of the labouring classes succumb? It was, surely, through defects of habit. It was because there were 'wasters'. Who became tubercular? It was those dissolute people who wasted their surplus wealth on luxuries such as tobacco and alcohol, leaving nothing for proper nutrition and domestic arrangements. It was the wastrels who dissipated their energies on irregular sexuality, above all, masturbation (many doctors claimed that tuberculosis and syphilis were related diseases, producing similar wasting effects). It was idlers. It was all those loafers who lacked moral fibre, and were weak of mind, spirit and character, perhaps hysterically or nervously inclined. H. de Carle Woodcock, a member of the council of the National Association for the Prevention of Tuberculosis, did not mince his words. Tuberculosis would always be found, he judged, where there was 'sexual vice', the 'alcoholic habit' and 'moral dirt': 'tubercle attacks failures. It attacks the depressed, the alcoholic, the lunatic of all degrees'. It was, he concluded, 'in truth, a coarse, common disease, bred in foul breath, in dirt, in squalor ... The beautiful and the rich receive it from the unbeautiful poor'.

Of course, as these remarks show, nobody forgot that tuberculosis was also the disease of the rich. But, by the close of the last century, attention had shifted away from them: fewer of the affluent were dying, and many of the well-off tubercular were being comfortably catered for in new health resorts abroad, such as Davos in the Swiss Alps, which functioned essentially as the site of plush hotels, pampering clients with fine service and therapeutically rich food. Longstay patients were meant to benefit from rest, sunlight, and the pure air.

Rather, at the turn of the century, haunted by fears of national industrial and military decline, of the survival of the fittest, of degeneration and recidivism, attention switched down the social scale. It was a classic case of victim-blaming. The poor evidently squandered their health by their wastefulness. But what was to be done?

A drastic solution became popular, unique to TB. The sanatorium. By 1910, 41 public sanatoria had been built in England for working class sufferers, mostly extensions of Poor Law infirmaries and run through the Poor Law administration. A degree of coercion, overt or tacit, was involved in their administration. Tuberculosis had gone on the statute book as a notifiable disease; at the same time, and more positively, under the 1911 National Insurance Act provision was made for financial support, so-called 'sanatorium benefit'. Armed with such sticks and carrots, it was possible to persuade, or pressurise, sufferers to enter the sanatorium. And once there, patients found their lives strictly regulated; permission and passes were required to go beyond the perimeter, alcohol and tobacco were forbidden, and distinctive clothing sometimes had to be worn. Stringent internal

*Figure 7* The building of large sanatoria for working class TB sufferers – a particularly British development – was in full swing just after the First World War.

discipline was enforced. Unlike lunatics in an asylum or prisoners in a gaol, sanatorium patients did not formally lose their civil rights, but the TB colonies bore many similarities to such institutions.

Unlike the wealthy who basked comfortably in the Alpine valleys, poor inmates were forced to undergo strict rigours in the sanatorium. Pure air and sunshine were central to the regime; but in many, following the philosophy of Otto Walther at the Nordrach Sanatorium in the Black Forest, this was accompanied by prolonged and systematic exposure to the elements at full force. Contradicting Beddoes's emphasis upon warmth, windows were kept wide open – often there was no glass – and beds were wheeled onto verandahs all the months of the year. Visitors not infrequently reported patients soaking wet, chilled to the bone, or with snow on the counterpane, an effective, as well as picturesque, way to shock patients out of their doubtless deleterious preference for the fug to which they had been accustomed at home.

Not least, the sanatorium affirmed faith in the therapeutic value of work. This often went beyond the commendable and constructive emphasis upon cultivating craft skills which was evident, say, at the Papworth Village Settlement, under the direction of Sir Pendrill Varrier-Jones. It became what can only be called a religion of hard forced labour, rationalised by some physicians, such as Marcus Paterson, the martinet superintendent of Frimley sanatorium (where there was no heating at all), through the auto-inoculation theory – the claim that the tubercular patient, by dint of demanding, outdoor, physical work – chopping trees, setting concrete, planting trees in the snow, etc. – generated metabolic reactions that would heal the disease. Paterson reported in 1908 that (to quote Bryder)

> the patients had dug, manured, and sown over an acre of grass, excavated for the walls of a new reservoir to hold 500,000 gallons of rain water, mixed and laid 650 tons of concrete, made most of the paths, laid the concrete walk to the dining hall, made a concrete subway 150 yards in length from the engine room to the kitchen, cleared a 20-foot 'fire zone' round the boundary, trenched and sifted about an acre of land, fetched and sifted gravel for the paths, made the terrace and rock garden round the tennis court by the medical officer's house, made a bank around the grounds, and felled and cut into firewood about 100 trees.

Patients who weren't frozen to death were thus in danger of being worked to death. The expressed opinions of men such as Paterson leave no doubt, however, that they believed that such labour – as in the workhouse and the asylum – would equally build moral fibre, instil discipline, and return patients to the outside world instilled with the right work ethic to prevent them once again from wasting their substance, to say nothing of tax-payers' money.

Those who returned, that is. For – though this was almost never said in so many words – the subtext of the sanatorium was that it was an institution from which it was expected that many would never return (three-quarters were dead within five years of entering the sanatorium). Nominally preventive or therapeutic, it was often terminal, a convenient dustbin where the dying could be despatched so they would never again infect family, neighbours and workmates. Surreptitiously, the sanatorium functioned as a segregative institution, protecting society from the degenerates within. Alongside the population of over 100 000 lunatics confined in asylums, to say nothing of idiots' asylums, gaols, reformatories and borstals, the sanatorium is a symptom of a society feeling desperately threatened by the moral contagion of the unbeautiful poor. No wonder Colindale was popularly known as Coffindale, and Grove Park as Grave Park.

## CONCLUSION

My theme has been a medical catastrophe; and I have been suggesting that the most serious long-term killer in Western civilisation – tuberculosis – drew to it a language, a cosmology, of moral penalisation. First, in the early modern era, consumption became a metaphor for all the traditional Classical and Christian evils of excess that endangered reason and order. By 1800, concern had shifted to worries about the civilising process itself, and the diseases of civilisation: consumption had become an emblem of the inner weakness of ruling orders grown too effeminate, needing more Spartan steel in their souls. Finally, by 1900, consumption was primarily a telltale sign of social disintegration, of class conflict, precipitated by the degenerate great unwashed, the wasters and the wasting, who needed to be ceremonially cast out of society and re-educated in the salutary habits of work. Disease catalysed perceptions of imminent social disaster. The history of TB helps trace the lineage of respectable fears.

## FURTHER READING

Sontag, Susan, *Illness as Metaphor*, New York: Farrar, Straus & Giroux, 1978; London: Allen Lane, 1979 and *AIDS as Metaphor*, Harmondsworth: Allen Lane, 1989, give a provocative introduction to the interplay of medical and moral fears.

Dubos, René and Dubos, Jean, *The White Plague. Tuberculosis, Man and Society*, London: Gollancz, 1953, is the best history of tuberculosis.

Bryder, L., *Below the Magic Mountain. A Social History of Tuberculosis in Twentieth-Century Britain*, Oxford: Clarendon Press, 1988, surveys the sanatorium.

Thomas Beddoes's views may be found in *Essay on the Causes, Early Signs and Prevention of Pulmonary Consumption for the Use of Parents and Preceptors*, Bristol:

Biggs and Cottle, 1799; *Hygëia: Or Essays Moral and Medical, on the Causes Affecting the Personal State of our Middling and Affluent Classes*, 3 vols., Bristol: J. Mills, 1802; and *Observations on the Medical and Domestic Management of the Consumptive: on the Powers of Digitalis purpurea; and on the Cure of Scrophula*, New York: O. Penniman, 1803.

More detailed and footnoted treatments of many of the issues raised in this chapter are contained in Porter, Roy, 'Civilization and Disease: Medical Ideology in the Enlightenment', in *Culture, Politics and Society in Britain 1660–1800*, ed. J. Black and J. Gregory, pp. 154–83, Manchester: Manchester University Press, 1991; *idem*, 'Consumption: Disease of the Consumer Society?', in *Consumption and the World of Goods*, ed. John Brewer and Roy Porter, London: Routledge, 1991; and *idem*, 'Bodies of Thought: Thoughts about the Body in Eighteenth Century England', in *Interpretation and Cultural History*, ed. J. Pittock Wesson and Andrew Wear, pp. 82–108, London: Macmillan, 1990.

## NOTES ON CONTRIBUTORS

GEOFFREY LLOYD is Professor of Ancient Philosophy and Science in the University of Cambridge and Master of Darwin College. Educated at Charterhouse and King's College, Cambridge, he has taught at Cambridge since 1958. He has lectured extensively in Europe, North America, Japan and China. His most recent books include *Demystifying Mentalities* (1990) and *Methods and Problems in Greek Science* (1991).

ROBERT P KIRSHNER is Professor of Astronomy at Harvard University and Chairman of the Astronomy Department. His work centres on observational extragalactic astronomy including supernovae, supernova remnants, galaxies and the large-scale distribution of matter. Graduating from Harvard College in 1970, he received his PhD at Caltech, and worked at the Kitt Peak National Observatory and the University of Michigan before moving to Harvard in 1985. Kirshner is Principal Investigator for a programme of supernova studies with the Hubble Space Telescope, and an active user of ground-based telescopes. He is the author of over 160 professional publications and has also written for *Scientific American*, *Natural History*, and the *National Geographic* magazine. He is the author of a chapter on the extragalactic distance scale in the recent volume on modern cosmology, *Bubbles, Bumps and Voids in Time* edited by James Cornell (Cambridge University Press).

WALTER ALVAREZ received his PhD in geology from Princeton University in 1967. His research has taken him to Colombia, the Netherlands, Libya, Corsica, Sardinia, Tunisia, Cyprus and Soviet Central Asia, as well as Italy, where his work on the Cretaceous-Tertiary boundary began. He is now Professor of Geology at the University of California, Berkeley, and an honorary citizen of Piobbico, in the Marche Region of Italy.

FRANK ASARO is a nuclear chemist in the Lawrence Berkeley Laboratory in the University of California at Berkeley. He has collaborated with Walter Alvarez for over 10 years on research into the impact hypothesis of mass extinctions.

*MARTIN RUDWICK* taught geology at Cambridge before turning to the history of science, which he has taught there and in Amsterdam and Princeton. He is now Professor of History, and a member of the Science Studies Programme, at the University of California, San Diego. His historical research is centred on the emerging sense of the vast scale of prehuman time and history in the decades around 1800.

*SIR CHRISTOPHER ZEEMAN* is a mathematician specialising in topology and dynamical systems. He has written a book on catastrophe theory as a method of mathematical modelling, and has pioneered its application. He was educated at Cambridge and was a Fellow of Gonville & Caius College. He then founded and directed for 24 years the Mathematics Research Centre at the University of Warwick. He is now Principal of Hertford College, Oxford and Gresham Professor of Geometry.

*CLAUDIO VITA-FINZI* was born in Sydney (Australia) and educated in Argentina and England. He took his BA and PhD in Cambridge and, after a spell as Research Fellow at St John's College, moved to University College London, where he now teaches in the Department of Geological Sciences. His research deals with recent earth movements and their bearing on crustal dynamics. He is the author of *The Mediterranean Valleys* (Cambridge University Press, 1969), *Recent Earth History* (Macmillan, 1973), *Archaeological Sites* (Thames & Hudson, 1978) and *Recent Earth Movements* (Academic Press, 1986).

*NICHOLAS J COOK* is the Director of the Geotechnics and Structures Group at the Building Research Establishment, and Special Professor in the Department of Civil Engineering at the University of Nottingham. He joined BRE in 1973 to design and build a boundary layer wind tunnel. Between 1973 and 1989 he developed modelling techniques and assessment methods, which are now standard practice worldwide. Recently he was responsible for the design and construction of the new boundary layer wind tunnel at BRE, which is the most advanced building aerodynamics wind tunnel in the world. He is the author of two text books for building designers on wind loading of buildings.

*PETER GARNSEY* is Reader in Ancient History at the University of Cambridge and Fellow of Jesus College. His main interests lie in ancient and comparative social and economic history, with special reference to famine and malnutrition. His most recent book is *Famine and Food Supply in the Graeco-Roman World: Responses to Risk and Crisis* (Cambridge University Press, 1988).

*ROY PORTER* is Senior Lecturer in the social history of medicine at the Wellcome Institute for the History of Medicine, London. He is currently working on the history of hysteria. Recent books include *Mind Forg'd Manacles. Madness in England from the Restoration to the Regency* (Athlone, 1987); *A Social History of Madness* (Weidenfeld and Nicolson, 1987); *In Sickness and in Health. The British Experience, 1650–1850* (Fourth Estate, 1988); *Patient's Progress* (Polity, 1989) – these last two co-authored with Dorothy Porter; and *Health for Sale. Quackery in England 1660–1850* (Manchester University Press, 1989).

*JANINE BOURRIAU*, editor of the volume, is a Fellow of Darwin College and of the McDonald Institute for Archaeological Research in Cambridge. Her current work is concerned with the archaeology of Egypt during the second millennium BC. She has worked in the Metropolitan Museum of Art in New York and the Fitzwilliam Museum, Cambridge and her most recent book is *Pharaohs and Mortals: Egyptian Art in the Middle Kingdom* (Cambridge University Press, 1988).

## *Acknowledgments*

### CHAPTER 1

*Figure 1* Harvard College Observatory photograph; *Figures 2, 3* European Southern Observatory photograph; *Figure 6* From Phillip Flower, *Understanding the Universe*, St Paul, Minn: West Publishing, 1990; *Figure 7* From the *Los Angeles Times*, 19 January 1934. Courtesy of the Associated Press; *Figure 8* Photo by Floyd Clark, *Engineering and Science Magazine*, California Institute of Technology; *Figure 9* Photograph from the National Radio Astronomy Observatory, Charlottesville, Virginia.

### CHAPTER 2

Walter Alvarez's article is based on an earlier article, 'An Extraterrestrial Impact,' by Walter Alvarez and Frank Asaro. Copyright © 1990 by Scientific American, Inc. All rights reserved. *Figure 1* From 'What Caused the Mass Extinction?' Copyright © 1990 by Scientific American, Inc. All rights reserved.

### CHAPTER 3

*Figure 1* Original drawing courtesy Pauline Dear, Ithaca, New York; *Figure 2* From De la Beche's 1830 field notebook, reproduced with permission from the British Geological Survey; *Figure 3* With permission from Scripps Institution of Oceanography, La Jolla, California; *Figure 4* With permission from the Geological Museum, London; *Figure 5* From De la Beche's 1830 field notebook, reproduced with permission from the British Geological Survey; *Figures 6–7* Prints supplied by the author; *Figure 8* From Jean de Charpentier, *Essai sur les Glaciers*, 1841, pl. no 1, reproduced by permission of the Syndics of Cambridge University Library; *Figure 9* From Louis Agassiz, *Etudes sur les Glaciers*, 1840, pl. 17, reproduced by permission of the Syndics of Cambridge University Library.

### CHAPTER 5

Claudio Vita-Finzi thanks Janet Baker for the figures, Mike Gray for the prints and Derek Banthorpe for his perceptive comments on the manuscript. Figure 3 was plotted by Steve Kaye; Figure 7 is based on work by Graham Yielding and others; earthquakes and rupture zones in Figure 8 on publications by H. Kanamori; and Figure 17 on publications by Roger Bilham.

## CHAPTER 6

Reproduced by permission of the Controller, HMSO: British Crown Copyright, 1990.

## CHAPTER 7

*Figure 1* Louvre E17381/photo by Dominic Rathbone; *Figure 2* From G. Pinto, *Il Libro del Biadaiolo: Carestie e Annona a Firenze dalla meta del 200 al 1348*, Firenze, 1978; *Figures 3–8* Fourteenth century miniatures from the R. Medicean Laurentian Library, Florence; *Figure 9* (*a*) From B. Martchenko and O. Woropay, *La famine génocide en Ukraine 1932–33*, Paris: Publications de l'Est Européan, 1983. (*b*) Information compiled from S. G. Wheatcroft, 'More light on the scale of repression and excess mortality in the Soviet Union in the 1930s', *Soviet Studies*, 42ii, 1990, p. 361; *Figure 10* Cited in E. M. Crawford, ed., *Famine, the Irish Experience, 900–1900: Subsistence Crises and Famine in Ireland*, Edinburgh, 1989; *Figure 11* Extracts from Lansdowne manuscripts, British Museum No. 48, f.128. Cited in E. M. Leonard, *The Early History of English Poor Relief*, London, 1900 (reprinted 1965).

## CHAPTER 8

Photographs courtesy of the Wellcome Institute of Medicine Library, London; *Figure 1* From Burnet, C. E. M. P., *Encyclopédie par l'image Pasteur*, Hachette, 1926; *Figure 2* 'The two greatest men in England' by 'CW', S. W. Fores, 1806; *Figure 3* From 'A Fellow of the College' [Peter Shaw], *The juice of the grape: or wine preferable to water*, W. Lewis, 1724; *Figure 4* From *Onania; or the heinous sin of self-pollution* (6th edn), T. Crouch, 1722; *Figure 5* From Morton, R., *Essay of consumptions*, London, 1834; *Figure 6* Wash drawing by F. C. Dickinson, 1901; *Figure 7* Colindale Hospital and King George V Sanatorium from Powell, A., *The Metropolitan Asylums Board and its work, 1867–1930*, London, 1930, figs. p. 58.

# INDEX

Illustrations are indicated by italics

0540–69.3, pulsar, 21

acid rain, and extinction, 45
alcohol, and health, 185–7, *186*
allopatric speciation, 95–6; *and see*
   speciation
amino acids, extraterrestrial, 50
ammonites, extinction of, 29–31, *32*
Anatolia (Turkey), famine (1560–1620),
   175
animal behaviour, and earthquakes,
   114–15
Athens (Greece), earthquakes, 121
Atlantis, earthquakes, 121
atmospheric boundary layer, 135–6

Bangladesh, famine (1974–5), 164
BCG vaccine, *181*, 181
Beddoes, Thomas, 191; on tuberculosis,
   191–6
Bengal (India), famine (1943), 164, 176
bifurcation, 84, *85*, 96
Big Bang, 6
black holes, 22
body wave magnitude ($m_b$), 108
Brighton Chain Pier, 128
Buckland, William, and catastrophism,
   74–7, 79–80
Building Research Establishment, wind
   data, 136
buildings: and earthquakes, 124, *125*;

storm damage, 127–8, 130–1, 137–44;
   and wind resistance, 127–8, 130, 134–7

cachexies, eighteenth century, 187
canalisation, 86, 88–9, 98–9, *99, 100*
Caravaca (Spain), K–T boundary, 33, 41
Cassiopeia A, *26*
catastrophe theory, and evolution, 83–101
catastrophism, 52–4, 60–1, 67–74, 80–2
Charleston (USA), hurricane damage,
   140–4, *141, 142, 143*
Charpentier, Jean de, and glacialism, *78,*
   78–9
Cheyne, George, 188–9; on health, 189–90
Chile, earthquakes, *117, 118*, 118; 1835
   earthquake, 116; 1960 earthquake, 108,
   *116*
China: 1976 earthquake, 103, 122; famine,
   147, 177; 1959–61 famine, 164, 170
Coalinga (USA), earthquake (1983), 118
cobalt, and radioactive decay, 23
comet storms, 47, 49–50
companion star hypothesis, 47–8
consumption: causes of, 187–8; symptoms,
   180; *and see* tuberculosis
consumption (economic), and disease,
   182–3
Coriolis effect, 129
cosmic rays, 25–7
Courtillot, Vincent, and extinction
   theories, 36–7
crude death rate, famine, 164; *and see*
   death tolls

209